A New Foundation for Representation in Cognitive and Brain Science

Springer Series in Cognitive and Neural Systems

Volume 7

Series Editor
Vassilis Cutsuridis
Boston University, Boston, MA, USA

For further volumes:
http://www.springer.com/series/8572

Jaime Gómez-Ramirez

A New Foundation for Representation in Cognitive and Brain Science

Category Theory and the Hippocampus

Springer

Jaime Gómez-Ramirez
Departamento de Automática
Universidad Politécnica de Madrid
Madrid, Spain

ISBN 978-94-007-7737-8 ISBN 978-94-007-7738-5 (eBook)
DOI 10.1007/978-94-007-7738-5
Springer Dordrecht Heidelberg New York London

Library of Congress Control Number: 2013953932

Printed on acid-free paper

Springer is part of Springer Science+Business Media (www.springer.com)

To my wife Sarah for her tireless support, love and uncanny feminine mystique

To our son Bertrand for having taught *me more that I have him*

Preface

> Home is where one starts from. As we grow older The world becomes stranger, the pattern more complicated Of dead and living. Not the intense moment Isolated, with no before and after, But a lifetime burning in every moment And not the lifetime of one man only But of old stones that cannot be deciphered. T.S. Elliot [1]

Manichaeanism or explaining facts based on two opposite poles is a habitual practice in politics and the media today. With this kind of explanation, we always find two clearly separated sides, one is good and the other bad. It goes without saying that the Manichaean placed himself in the good side. Following Jeannerod [2], this polemic way of seeing the world is hardwired in the human brain, so as this book has been entirely conceived of and written by a human – no hermeneutical algorithm à la Sokal here [3] – I ask everyone to forgive me for starting with such a dichotomic differentiation to make my point. In a Manichaean stance, there are two ways of modelling cognitive systems. One is meant to demonstrate that we understand some particular aspects of the system, and the other tries to discover new structural and functional relationships in the system. The former allows numerical simulation, while the latter attempts to capture the organising principles underlying the behaviour of the system. In conclusion, the former is data driven, while the second is driven by the creative act of the modeller and ultimately endorsed by the experimental data. This book aspires to be considered under this second conception, and essentially is an attempt to help in the transition from a descriptive cognitive science to a strongly mathematical one in either its methods or concepts. Such a transformation has just started and seemingly will keep going on.

The moon-earth system can give us a nice historical introduction to the problem that is at the core of this book. The Ptolemaic's system was able to predict the position of the planets, so it was a good model in simulating data which were scarce at that time. The trouble came when the data set became abundant; here the experimentalist Tycho Brahe comes to mind. In this new scenario, the Ptolemaic model, in order to keep pace with the new influx of data, became more and more complex, even baroque. Copernicus and Kepler's revolution means the substitution of the Ptolemaic model that was unmanageable by a more simple one. Newton later came up with the law of universal attraction of two bodies described in a succinct mathematical form and empirically verified with the available data on the moon-earth relative positions. The conclusion we can take from this is as follows. The Ptolemaic model was able to deal, until a certain point, with data, but it was necessary for Copernicus and especially Newton to strive for mathematical structure in order to discover the underlying principles or laws that govern the attraction of any two bodies, moon and earth included. Thus, Newton happily solved the two-body problem analytically, but if we include just a single body to our system which now has three bodies, finding a solution becomes impossible! This was true for 300 years, until Poincaré's breakthrough in rethinking the problem with the question, "Is the solar system *stable* forever?" Poincaré, by focusing on stability – a systemic property that cannot be found in the individual planets but in the system as a whole – was allowed to succeed where many brilliant minds before him had failed. The limitations in predicting the position at any time of any of the three celestial bodies – sun, moon and earth – were only clearly understood when the emphasis shifted from the correlation between the properties of individual entities to looking into the organisational principles of the entire system.

There is a radical difference between biological and physical systems modelling; while physics deals with stable objects (of course there are exceptions like the radioactive substances), in biology, systems are always getting ready to change to a new state. Biological systems have a certain purposiveness, that is, an arrow in the temporal direction. In physics, if we represent the transition between two states with the mapping, $A \rightarrow B$, the uncertainty is usually in the final state, B, and particular boundary conditions for A are imposed in order to get a unique solution for B. On the other hand, in biology the uncertainty is in the arrow itself. Thus, we can predict that from A the system will evolve to B, but there is a multiplicity of ways to do so. For example, the output of gene expression may vary substantially depending on the different cells in which it occurs. In precise mathematical terms, gene expression is a stochastic process.

D'Arcy Thompson, Nicolas Rashevsky and Alan Turing were pioneers in the building of a theoretical framework that could put biology and physics at a similar epistemic level. The elegant prose and harmonious sense of aesthetic beauty of the drawings in Thompson's *On Growth and Form* [4] should not distract us from the fact that his examples only apply to objects with symmetries and clear morphological regularities. Thus, Thompson's theory cannot be universal. For example, the three-dimensional structure of proteins may have a total lack of symmetries. This was discovered in 1958, 51 years after the first edition of Thompson's book, by John Kendrew [5] who determined the first atomic structure of the protein myoglobin.

Function is a word with multifold interpretations. The precise and colourful three-dimensional structure of the proteins now obtained in laboratories does not tell us its function inside the cell. In engineering, the function of a system is not problematic because it is the engineer who gives to the system its specific function, e.g. the function of an end effector in a manipulator arm is not to grab things, it is rather the engineer who confers such power to an object endowed with properties that make it appropriate for the task. In sharp contrast with this, function in biology is the result of evolution. Thus, the function of myoglobin is to store oxygen in muscle cells and that of protease is to decompose other proteins.

The modern field of mathematical or theoretical biology started with Nicolas Rashevsky's approach to biological processes using the methodology and goals of mathematical physics. Rashevsky's relational biology focused on the search of general principles in biology. However, the systemic approach proposed by Rashevsky, Rosen and others failed to be conveniently explored. The emphasis of the early theoretical biologists on emergent properties in entire organisms, e.g. life itself, was far removed from the reductionist methodological approach championed by twentieth-century biology. Furthermore, biology researchers were confronted with a mathematical language for which they lacked formal training [6]. As a result, the beginnings of theoretical biology were disregarded as abstractions detached from the work of the experimental biologists, which was mainly driven by the study of gene expression. Ironically, the obsolescence of the concept of gene [7] is bringing back to the fore the same tenets that were set forth by the early theoretical biologists, now under different designations. For example, systems biology focuses on the study of the organisational principles that mediate in biological function using mathematical methods.

Francis Crick is known for his discovery of DNA, but this achievement is due in part to his theoretical recognition that in order to study the function of a system, one must first determine its structure. But the structure that Crick had in mind cannot be the morphological configuration that is provided, for example, in the static picture of X-ray crystallography that Kendrew obtained for the myoglobin. If that were the case, the mapping $Genotype \rightarrow Phenotype$ would have been already resolved. Of course, we are far from that scenario. The structure must encompass the dynamic relationship of the components that preconfigure the system, as well as its evolving patterns, and preferably has to be defined in mathematical terms. The strictly linear causal scheme $Genotype \rightarrow Phenotype$ must be replaced by another that encapsulates the multilevel interactions of the system's components, together with the changes that the environment produces in the system and the regulatory mechanisms elicited to cope with this variable form of interaction. Bearing this in mind, it could be argued that in order to understand the structure and function of the neural system, a similar approach to functional genomics should be pursued. This kind of linear reasoning is, of course, wide of the mark because the brain is an infinitely more complex system than a DNA molecule. It might be emphasised that a DNA molecule has to be simple; it is the origin of life, while the brain is the extremely elaborated result of a long history of evolution. The devil is in every detail. But we can always get rid of the devil, by choosing a system not so overwhelmingly complex as the human brain. For example, a simple organism like the soil worm Caenorhabditis elegans has been cracked down, and a complete map of the mere 900 cells that this organism has, of which 302 are nerve cells, is available [8].

How does the neural system encode mental objects, in particular spatial representation and memory? This is the issue that this book addresses, and in doing so, it borrows concepts from the mathematical theory of categories. In the end, brain function and behaviour depend upon the interaction of elements – proteins, cells, networks, brains, etc. The neural and cognitive sciences deal with complex components (assemblies of cells, brains) and out of necessity will follow a mathematical strand. Science moves in little steps, but also makes its progress with revolutionary discoveries and concepts that sweep away whole and entire edifices of thinking and replace them with new theories that explain more with less. However, there is a constant in this march, the strive for mathematisation and unification. The dynamic nature of biological organisms has to be understood, not in the well-known evolutionary sense of organisms changing over time but in terms of the persistence of patterns at the different levels of description, from molecular transcriptional networks up to cells, tissues, organisms and ecosystems. When Schöringer wrote in 1944 *What is life?* [9], the biology at that time could be explained without mathematics. He argued that this was not because biology was simpler than physics or chemistry, but quite contrary to this, biology was too complex to be mathematised. In 2013, it seems foolish to deal with such complexity without mathematics. The new science of complex systems has exploded in the last 50 years. Network science, non-linear dynamics, chaos theory, statistical physics ... are now part of the academic curriculum. However, we must be prudent when talking about paradigm shifts or Kuhnean revolutions. As the evolutionary biologist Ernst Mayr has observed, in biology there are no scientific revolutions, at least not in the Kuhnean interpretation [10]. This is, Mayr argues, because in order to have revolutions or paradigm shifts, we need first to have universal laws to be surpassed by new ones, and in biology we lack the kind of the universal laws that exist in physics.

Before I conclude, I have to confess a final weakness here, which is that of being a dilettante. It is not rhetorical apology to acknowledge the limitations and difficulties that lie ahead this new approach. But it may be the case that it is in the hidden road, rather than the well-signalled path, where the kind of breakthroughs needed might be found.

Madrid, Spain Jaime Gómez-Ramirez

Contents

List of Figures

List of Tables

Chapter 1
Research Tools and Paradigms

1.1 Introduction

Biology in the twenty-first century will be for Mathematics what Physics was in nineteenth and twentieth centuries. This is a well accepted belief among scientists with mathematical training. Some physicists can go even further and deplore the pre-Newtonian stage in which biology and in particular neuroscience find themselves [25]. Neuroscience is a data-rich field that needs for a theoretical framework that guides the model building and simulation processes. Biology, in opposition to physics which is quantitative and explanatory, may be perceived as a descriptive and qualitative field. However, to state that biology is descriptive and qualitative, while physics is mathematical and quantitative is a dichotomy too simplistic to be true (as all dichotomies usually are). Undoubtedly, biology today is quantitative in either its models and results. It is easily noticeable that in top scientific journals such as *Nature* or *Science*, though they are not devoted to any particular field, the studies in the field of biology, specifically molecular biology, are the overwhelming majority of the total of articles. The very dissimilar rate of production that these two prestigious publications display between disciplines other than biology is such, that one starts wondering if the non-life science community has anything left important to say. But the actual reason for this must be found in the scarcity of mathematized universal principles in biology compared to, for example, physics. Physics is built on formal theories embedded in universal laws, rendering particulars and details unnecessary. In biology, on the other hand, particular cases are relevant. Furthermore, variability, non linearity, noise or high dimensionality are biological features hardly generalizable in the standard mathematical forms [26].

The pace at which new experiments and data are being produced in the life sciences is overwhelming for the specialist and disconcerting for the non-specialist. In what follows I address the following two questions: why has biology not yet attained the same status of physics in terms of rigour and predictability? and why do we still lack a unified theory of cognition? We must stay far from far overarching responses, instead we must foster a new way of understanding complex biological

J. Gómez-Ramirez, *A New Foundation for Representation in Cognitive and Brain Science*, Springer Series in Cognitive and Neural Systems 7, DOI 10.1007/978-94-007-7738-5_1,

organisations. Neuroscience is at the very beginning of a long way of discovery that may be conducive for a satisfactory understanding of the mechanisms that govern cognitive functions such as learning or memory. This state of affairs has been succinctly stated by Walter J. Freeman "Brain science is still in adolescence [27]". A conceptual breakthrough in biology is imperative, as a result, a new and deeper insight to the organisational principles of living systems will emerge. The mathematical articulation of a theoretical framework will encourage a new way of thinking in life sciences.

This chapter explores the idea that cognitive and brain science needs a transition from a purely descriptive to an axiomatic one. The following sections present a re-examination of this claim, and it does so by, on the one hand, relating it to the actual historical development of physics from the time of Newton and, on the other hand, describing what brain science might learn from this development to help put it on a more fundamental and logically sound footing, if anything. This survey may allow for a more efficient discovery and development of the basic theory of brain science.

1.2 Mathematics as a Language and as a Modelling Tool

Already in the first paragraph of the popular book "What is life?", Schrödinger announced that mathematics would hardly be utilized in biology [9]. In Schrödinger's view life was too complex to be described in mathematical terms. This idea, since the book was written in 1944, has prevailed as a credo that does not need to be discussed, especially if one wants to avoid being called a reductionist with a flawed vision about what life is. Rudolf Carnap, the Vienna Circle's philosopher, was the champion of physicalism whose main tenet is that all sciences, including human sciences, were reducible to physics, because every predicate of psychology, biology ... may be translated into sentences within the physical domain [28]. This extreme reductionism of Carnap is untenable. It is self-evident that the universal laws of physics must necessarily apply to living systems, thus physics and chemistry are, of course, relevant in biology. But, this only proves that nature is ordered and probably hierarchically structured, that being so, a predicate of level i is irrelevant, though true, at superior the levels, $i + 1, i + 2 \ldots$. For example, the chemical composition of the ink with which the book is printed, is completely irrelevant for the literary critic, just as the ideal gas law is of no interest for a researcher in population genetics.

In a similar fashion as physicists do in abiotic domains, system biologists have started studying the general physical principles at work in biology. It might be remarked that the quest for a system-level understanding is far from new, physiology or the study of living systems in an integrative way dates from Ancient Greece. The increasing interest in systems biology must be found in its capacity to use a variety of mathematical models and computational tools directed to the understanding of biological processes, normal and pathological, anchored in molecular data [29, 30]. In essence, this is the engineering approach which does not commit itself with any particular level of description or set of details, rather it focuses in the functional level of description that aims to capture the general principles that explain why the system behaves in the way it does. In physics or chemistry, given a system, one may conjecture about its structure. It should be possible to make predictions, experimentally tested, about the future states of the system. But in biology, for example when we study proteins, knowing the 3D structure of the protein, is clearly not enough to understand its role played inside the cell. If we modify an enzyme expecting to block its function, another protein could take that role [31]. This is due to redundancy, one of the key organisational principles in biological systems.

1.3 The Development of Physical Theories

Physics is often used as the prime example of a scientific area of study, that of "hard" matter, with the assumption that physics was developed logically, based on formal theories, embedded in universal laws. I suggest that this is not a complete picture of the area, and that it could lead researchers in other fields, seduced by the supposed crystal-clear logic of the situation, into possibly expecting too much for their theories. It ought to be mention that physics is far from being a pristine and healthy scientific discipline where ambiguity and paradoxes are absent. It needs just to bring the physicist to the domains of the very large and the very small, to see that the concepts and measurements used in their theories have a non-unique interpretation, as it happens in social science. Try to mimic the successes in physics in neuroscience, by assuming a privileged relationship of the physical modes of explanation and reality, could possibly lead away from the basic problems being faced in an area, such as brain science, where there is some confusion about what the theoretical basis of brain science should look like.

Physics has made enormous strides in our understanding of matter, leading to the construction of the Large Hadron Collider (LHC). But there is still considerable confusion in modern physics as to what is being looked for at this vast experimental apparatus, with some considering that supersymmetry, higher dimensions, the Higgs boson and even possibly superstrings may be glimpsed as the energy of collision is turned up and all the event pictures are laboriously investigated. Other physicists assume that nothing new will be seen.

Physics got to this stage of creating the LHC by the usual scientific process of

1. A large step taken by powerful enough theory being proposed to explain a whole host of past data,
2. The process of mopping up further areas from which data can be collected, all this being explained as part of the theory or its extensions
3. Recalcitrant data being discovered which blows apart the initial "beautiful theory", being resistant to explanation by any of a whole host of variants of the theory

Table 1.1 Mathematical approaches in physical theories

Theory	Mathematical structure	Dates
3 laws of motion	Dynamical effects in n-dimensional flat space leading to dynamical systems theory and theory of complexity	1500–1600
Universal Inverse Square Law of Gravity	Added to Newton's three laws of motion to explain planetary orbits	Ditto
Quantum Mechanics	Hilbert spaces and operators on them	1915, 1923–1930
Special Relativity	Modified metric in 4-dimensional space-time	1905
General Relativity	Riemann surfaces in 4-dimensional space-time	1935–1940
Quantum Gauge Theories	Ditto, but now with gauge properties	1965
Superstring Quantum Field Theories	Ditto	1986
M-Theory	Not known fully	1990s

4. A powerful new theory being proposed which begins to explain the recalcitrant data,
5. Consolidation of the new theory, both by it being shown to explain further recalcitrant data to the old one as well as being able to allow derivation of the old theory in a restricted area of environmental conditions (by so-called bridging laws).

The steps 1–5 above are all well-known. Thus, the transition from Newton's three classical laws of motion was achieved by the invention of quantum or wave mechanics. The bridging laws arose by taking large numbers of particles or other reasonably well defined limits (although still with controversy). Quantum mechanics has been very well tested in many environments and undoubtedly is now completely accepted.

The message to be learnt from this brief note on the underlying nature of matter over the last three centuries is that of "hit and miss". The initial suggestion of a basic theory of great power and beauty is a hit, and may remain so for some years if not centuries (as in Newton's case). But then new areas of exploration turn up the misses of the basic theory, so a new one has to be introduced. Such a process may be painful to those who had an emotional investment in the prior basic theory. They may well dig in their heels, and refuse to accept the new theory (as occurred for Newton and for Einstein, as well as by some who are still recalcitrant, despite the contrary evidence, such as Einstein, for the founders of quantum mechanics, Heisenberg, deBroglie, Schrodinger and Dirac). The historical development so observed is summarised in Table 1.1.

We notice from Table 1.1, that a wide range of mathematical approaches have been used in the various theories. However one crucial message to derive from the Table is that the two main branches of theories, those based on Special and General Relativity and those on Quantum Mechanics, have now been subsumed, at least theoretically, under the main umbrella of superstring theories. However there is no guarantee that this set of theories will prove satisfactory when new data arises from the LHC or even later from its successors. Thus, the main lessons are that as time progresses a variety of mathematical theories are necessary to formulate and evaluate that theory or range of theories supposedly fitting the latest range of data, but then those theories need to be changed to a quite different new set of theories to be able to proceed successfully. There seems to be no evidence of the existence of a final theory from which all of the understanding of matter could be derived. The additional point to be learnt is that of levels: as experimentalists dig ever deeper into the interior of the interior ... of matter, new theories, requiring new mathematics, arise.

There are two aspects that immediately arise in trying to improve the rigor and value of the mathematical analysis of the brain:

1. Are altogether new mathematical techniques needed, such as category theory or fibre bundle theory?
2. Is it of value to give support to more powerful techniques in branches of mathematics already being applied to the problems of the brain?

Undoubtedly the answer to both questions is a resounding "yes". But we should consider a comparison with the developments in theoretical physics. There we can recognise that the ultimate theoretical structure at which we have presently arrived is that of quantum field theory, across all the forces of nature. This is because we see that General Relativity has been folded into superstrings over the last quarter of a century, and this latter has been constructed on a quantum field theoretical basis since then. Indeed the theory of superstrings has been hailed in the manner it has because it is suspected as giving a theory fusing quantum theory and gravity. It is that problem – of combining gravity in a unified manner with the other forces of nature – that has confounded many attempts to achieve that aim since the 1940s. There is no equivalent situation for theories of the brain. If we regard the fundamental entities of any such theory as the neurons (furnished with their synapses and additional neuro-modulators where appropriate) then our basic theory must be that of such neurons. If we accept that neurons have an underlying stochastic manner of processing, caused by intrinsic synaptic noise and other sources of noise in fluctuations of thresholds, etc., then the basic mathematical structures are those of stochastic differential equations. We may attempt to take account of this by means of a continuous approximation of the distribution of neurons, so leading to continuum stochastic neural field theory.

One final feature of great importance for any comparison with brain science is that the theories above in Table 1.1 involved probing to ever smaller distances inside the fundamental units of matter, uncovering their components in the process. There is no guarantee that this process will ever cease, since there may indeed exist components of components of ... without end. Here, we defend the view that a possible way out of this conundrum, resides in adopting a systemic approach to study the brain as an open system with multiple levels of description, in which key concepts like causality or stability need to be carefully reconsidered under a new theoretical framework. The multiscale and stochastic nature of complex systems like the brain, and its emergent or systemic functions i.e., cognition have to be taken into account.

1.4 The Development of Brain Science

The question we can now pose is: what lesson can be learnt from this development of theoretical physics over the last three centuries, for brain science? To be able properly to answer this, we need also to give a brief survey of the development of brain science itself over the last 100 years or so, in order to see how the lessons we have stated above apply: no theory is yet the final theory, and each theory in the temporal sequence may require a new sort of mathematics. However, brains already exist at a much larger size than most of the objects involved in the theories of Table 1.2. Thus, certain processes in the very small in brains may need a quantum mechanical interpretation, but most processes appear to exist at the classical, Newtonian level. There is also the feature that the structures in brains are not only larger than those needing a quantum mechanical description but are mostly completely known, at least in outline.

Nerve cells have been known since the end of the nineteenth century. Ramon y Cajal, using a histological approach based on observations of stained cell structures, proved the discrete nature of the nervous system, that is, nerve cells, later called neurons, are separate entities that connect between them through axons. This is the core of the neuron doctrine that was disputed by Golgi, the co-recipient with Cajal of the Nobel Prize, and champion of the reticular theory of the brain. Interestingly, Golgi himself was who provided the experimental method, the Golgi stains, that allowed Cajal to demonstrate the existing discontinuity, at the termination of the axon or the dendrites, between cells. Significant achievements in brain anatomy followed. For example, Brodmann areas are known since about 1909. However, the origins of modern neuroscience must be found in the 1950s, when path-breaking

Table 1.2 Mathematical approaches in brain theories

Component investigated	Mathematical structure
Ion channels	Dynamical systems theory
Synapse	Ditto
Single nerve cell	3-d geometry, theory of Green's functions and non-linear dynamics, and stochastic equations
Multi unit activity (MUA)	Graded responses
Columnar activity	Dynamical systems, fibre bundles
Brain modules	Physics of phase transitions and synchronization; continuum neural field theory and fixed points of non-linear field mappings
Brain network of modules	Coupled dynamical systems and theory of coupled oscillators
Global brain	Behavioral responses

discoveries like neuromodulation on nerve cells, or the 1952 Hodgkin and Huxley model on the propagation of action potentials were established. So we can define a table similar to the previous Table now related to brain science. Here we only really need to go down to ionic channels, where quantum effects may even be neglected. So we are in the classical regime throughout. Table 1.2 indicates a large range of mathematical structures. These depend heavily on dynamical systems theory (notions of dynamical trajectories, attractors of a variety of sorts, attractor boundaries, etc.) as well as on stochastic dynamical systems. Fixed points of non-linear mappings are of relevance as possible phase transitions of synchronised neural activity and associated timing effects.

There is an important difference between the entries in Tables 1.1 and 1.2: in Table 1.1 the most relevant entities and associated theories are still evolving, whilst in Table 1.2 the main evolution is not of the basic entities but of their detailed properties. Focusing on the epistemological aspect of the problem, we can ask: Is there an universal basic theory, albeit one changing as the years pass due to new data, which we can recognise in brain science as clearly as we can in the fundamental physics of matter? The answer would appear to be yes, if entities at a certain level are considered. Thus, if we stop at single neurons then the basic theory is that of stochastic equations for the probabilistic firing rates of the coupled neurons. However each of the neurons itself has a specific geometry, so we need to include this by a compartmentalised model of the neuron, increasing the dynamical variables for each neuron depending on the number of compartments introduced. If we go down to the synapses then we have further dynamical equations for each synapse associated with the available and resting, etc. synaptic vesicles in the synapse. These variables must be added to those of each single neuron, since the dynamics of the latter is a separate process compounded of all the synaptic activity over the neuron in the recent past [32].

1.5 YAPS! Yet Another Paradigm Shift!

Understanding that biology is rich in proper names but poor in formal concepts is key. Indeed, biology textbooks abound with accurate descriptions of thousands of molecules, organelles and cells. In contrast, the same books give little account for the laws or general principles that govern the interactions between those molecules, organelles or cells. Mendel's law of inheritance is one, if not the only law, expressed in formal terms in biology. It ought to be noted that Mendel law is not universal but general, because it only applies to eukaryotic and sexuate organisms. This rarity of universal laws in biology marks a clear distinction with physics and poses questions that need to be taken into account.

Biology has historically focused on experiments, in order to a posteriori deduce theories and models underlying the phenomena under study. In physics, it is the theory that has lead its development and progress. For example, Newton built a model that captured the relationship between the moon and the earth in the movement of the former around the latter. Thus, through an abstraction Newton was able to understand the structure of the natural phenomena, i.e. the movement of bodies around each other. It is important to remark here that Newton's theoretical work precedes the recollection of data. As a matter of fact, he built the model not to fit the data; rather the mathematical abstraction that he constructed made it possible to understand the phenomenon, capturing its structure and putting it in mathematical terms [33]. There are of course exceptions to this preponderance of theory in physics, for example the discovery of Ray X and radioactivity that arose as unexpected experimental observations in the laboratory.

Here I share the view of the Oxford biologist Denis Noble, "the next stage in the development of biological science will be revolutionary in its conceptual foundations and strongly mathematical in its methods [34]". However, we must admit that meeting the expectations of the two traditions – the mathematical-logical and empirical-biological – is not a realistic goal, at least in the short term. The prime intellectual task of the future lies in constructing an appropriate theoretical framework for biology [35]. The task is ambitious and precaution is in order; we must first acknowledge the obstacles that we are facing here. Each scientific community uses its own language, the same word may mean very different things depending on the domain of discourse that is relevant for the occasion. By way of illustration, some examples of words with different meanings between biologists, mathematicians and physicists follow:

(i) To *Differentiate:* in physics is to find the slope of a function, in biology refers to a change in cell function.
(ii) To *Divide:* in mathematics is the operation that gives us the ratio of two numbers. For a biologist, division is the process of duplication by which a cell replicates its content and split into two, the characteristics transmitted by the mother cell are preserved in the duplicated cells. Division is a key process in biology because all cells, with the exceptions of neurons and cardiac cells, duplicate.
(iii) *Function:* in mathematics – set theory – a function is a relation between a domain and a codomain, each element of the original set or domain corresponds only one element of the target set or codomain. The most extended notation for function is $f : X \rightarrow Y$ where f is a function, X the domain and Y the codomain. In biology, molecules have associated functions, for example a certain molecule is said to be an enzyme because of its behaviour in the context of a particular network of chemical reactions, not because of any property it possesses independently of this network of reactions [36]. Thus, molecules have associated functions e.g., the function of the ribosome is to synthesize proteins and the function of a protease is to decompose other proteins. This interpretation does not apply in physics. In fact, a superconductor does not have a function, but properties. The function of a superconductor is given by the engineer in his strive to operate in optimal conditions of conductance. Using the words of John Hopfield, the creator of the associative neural networks, "The word function doesn't exist in physics, but physicists are going to have to learn about it" [37].

(iv) *Cell:* in mathematics – topology – means edges that connect vertices. The most common use of cell comes from physiology, cells are the smallest living things and they have all the properties of life, including reproduction, response to environmental signals, and metabolic cycle.

It might be emphasized that having identical words that express dissimilar concepts, depending on the context in which the word being used, is not intrinsically bad nor good. The trouble is that different scientific disciplines, when missing a common framework, can hardly understand each other. For Kuhn, such framework is possible within the context of evolutionary epistemology of theories [38, 39]. In contemporary academic science, interdisciplinary has become a buzz-word à la mode. Nevertheless, the preponderant traditional academic culture does not necessarily reward interdisciplinary, rather it is often viewed with suspicion or disdain. And perhaps not without reason, selling interdisciplinary research prior to build a common epistemological culture, may not seem an optimal strategy. A fruitful dialogue between different scientific communities must have first in place a common epistemic framework, preferentially expressed in mathematical language. Quoting the oncologist Robert Gatenby:

> Existing mathematical models may not be entirely correct. But they do represent the necessary next step beyond simple verbal reasoning and linear intuition. As in physics, understanding the complex, non-linear systems in cancer biology will require ongoing inter-disciplinary, interactive research in which mathematical models, informed by extant data and continuously revised by new information, guide experimental design and interpretation [40].

The same can be said about mind and brain. The cognitive and brain sciences must necessarily follow the systemic, integrative and mathematical based undertaking referred above. What is needed now in cognitive and brain sciences is to be able to be as successful and rigorous in putting together the pieces that reductionists were able to separate. Such an achievement requires an understanding of the brain at system level. The time is ripe to take a step back from the details and try to see the big picture.

1.6 Plan of the Book

I now proceed to give a chapter-by-chapter overview of the topics covered in this book. In this initial chapter we explored the question, why has biology not yet attained the same status of physics in terms of rigour and predictability? and whether we can expect a Kuhnean revolution in brain and cognitive sciences. In Chap. 2 different mathematical tools to model brain structure and dynamics are described. The microscopic, mesoscopic and macroscopic levels are covered in this review. The spirit of the chapter is to provide the reader with a mathematical toolbox to deal with nonlinear brain dynamics. In Chap. 3, category theory is introduced. I will provide a formal introduction to this mathematical theory, including applications in the domain of brain and cognitive science. A new and original theoretical framework in which different tools and approaches can be used in conjunction with one another, will naturally arises in this chapter. Chapter 4 studies crucial philosophical topics like formalisation, reductionism and emergence from a transversal and systemic perspective. It paves the way for the formulation of *structure* and *representation* in the following two chapters. Chapter 5 is devoted to *structured systems*. The concept of structure is rigorously explained here. In addition, I use category theory to introduce the concept of *cat-system*, which is a categorical formulation for *structured system*. In Chap. 6, I present a general framework for *representation* based on category theory. The main idea is to bring this mathematical formalism into the domain of neural representation of physical spaces, setting the basis for a theory of mental representation, able to relate empirical findings, uniting them into a sound theoretical corpus. Up to Chap. 6, we have treated the key issues of this book, i.e. structure and representation in a general setting. In the last part, the focus is shifted to the brain, in particular the hippocampus. The structure of the hippocampus, and the mechanisms by which mental representation and memories arise from its interactions will be conveniently covered. Chapter 7 explores two main issues. The first concern is to draw attention to the restricted, yet inconclusive quarrel between the different schools of brain organisation and functioning. The second concern is to show that networks can be seen as an abstraction of categories. By generalizing networks using category theory we can address questions and elaborate answers in a more fundamental way without waiving graph theoretic tools. We demonstrate that category theory can be successfully applied to network quantitative analysis, providing new metrics. Chapter 8 describes the effect in injecting the concepts of coproduct and colimit from category theory into the problem of place cell formation in the hippocampus. The hypothesis that will be explored here is that the colimit is the mathematical structure that allow us to encode the emergence of mental concepts related with space such as position, orientation or distance. The main point is that this theoretical postulate will shed light on how populations of grid cells contribute to the formation of one place cell. In Chap. 9 it is described a mathematical formalism based on the theory of categories for modeling declarative memory. The categorical concepts of product and pullback will be explored and mapped onto psychological structures related with semantic memory. Finally, in the Chap. 10 conclusions and future directions are discussed.

Chapter 2
State of the Art: Mathematical Approaches in Brain Science

2.1 Introduction

This chapter explores some of the most relevant mathematical structures used by brain researchers to unravel brain's structure, function and dynamics. It starts by looking into the concept of brain state. The concept of the state of a system is key in engineering and physics, but maybe it is a not a very well understood concept in other relevant fields of brain science like cognitive psychology. Scientific models and theories of the brain are empirically tested against signal measurements of brain activity. Undoubtedly, the methodology and technology used constrain the possible interpretations of the models. For example, single cell recording typically consists on time series of voltage values that represents spikes or action potentials. Spikes are characterised by two parameters: amplitude and duration. Recording a population of neurons, on the other hand, is far from trivial. The recorded signal is not a point process, rather it represents a summation of several events in the neighborhood of the electrode. Neuroscience constitutes a challenging and promising testing ground for developing our understanding of how the macroscopic level emerges from the interaction of large numbers of interacting components. The functioning of the brain at the high level of mental processing is, of course, expected to be the result of the co-operate action of a very large number of neurons.

For conciseness, we address here microscopic, mesoscopic and macroscopic brain levels. It might be noted that the traditional classification of hierarchical levels in biology – molecular, cellular, organisms, populations and ecosystems – is a purely methodological recourse and other classifications are perfectly possible. If we accept that there are different levels of description in the brain – synapses, neurons, columns, modules, etc. – each with their own dynamical systems and time flow, then we would expect that there are possibly different mathematical structures at each of the relevant levels (Fig. 2.1). A concise exploration of the different mathematical structures and how they apply to the microscopic, mesoscopic and macroscopic, are sketched in Sects. 2.3–2.5 respectively.

J. Gómez-Ramirez, *A New Foundation for Representation in Cognitive and Brain Science*, Springer Series in Cognitive and Neural Systems 7, DOI 10.1007/978-94-007-7738-5_2,

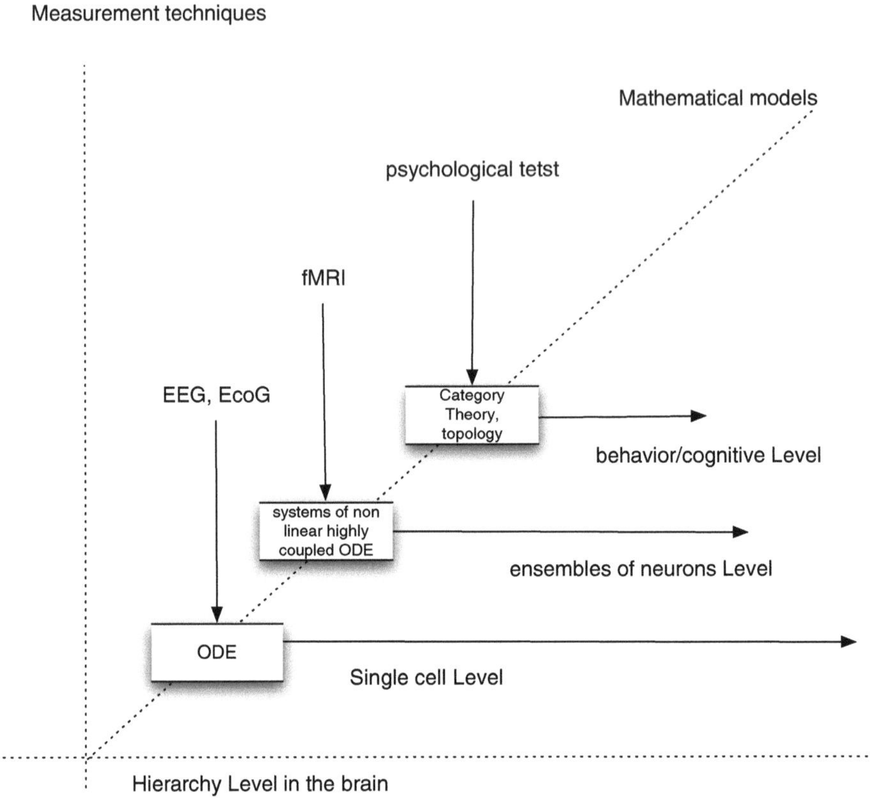

Fig. 2.1 Measurement techniques, brain hierarchy levels and mathematical models are related as the figure shows. An unique and all encompassing mathematical model of the brain cannot be expected. Note that Category theory can be used at multiple brain hierarchy levels

2.2 Brain State

The brain is a dynamical system of unprecedented complexity, its intricate organisation is the product of thousands of years of evolution and social interaction. The state space approach is used by engineers and physicists, and it roughly represents the system as the space defined by its states or configurations. Thus, the set of all possible trajectories or sequence of states defines the system's behavior. The brain is a dynamical system, which can always be studied using dynamical system theory (DST). DST is basically concerned with the study of the trajectories in the system's state space. This leads us to discuss the controversial notion of brain state. There are scholars, mainly philosophers who repudiate the idea of brain state. For example, quoting Polger "brain states ... we don't really even have a clue what such things are [41]". Bechtel and Mundale have argued that "the notion of a brain state is a philosopher's fiction [42]". But still, what brain states are and how can we discern them? Buzsaki defines brain state as a transient equilibrium condition, which contains aspects of past history that are useful for future use [43]. This definition emphasizes the inherent historical aspect of complex system. Complex systems, as Prigogine said, do not forget their initial conditions: they carry their history on their backs [44].

Synchrony, or more precisely the phase locking of the firing of the neurons may make help us identify brain states. For example, electrophysiological recordings in wondering rats show theta-modulated-gamma waves in the hippocampus [45]. However, the association between wave frequency pattern and functional state is not one-to-one. Indeed, in strictly electrophysiological terms, waking and paradoxical sleep are fundamentally identical states [46]. We need to be able to discern how perceptual information of the external world coming from very different sensorial apparatus is processed, integrated and finally reflected in the synchronous waves measured in electroencephalogram (EEG). Freeman proposed the analysis of the amplitude modulation as the way to unravel the perceptual content that is conveyed in the wave. Nunez quantitative theory of neocortical dynamics [47], defines brain state in the context of the theory, using psychologically based control parameters which also mediate in EEG frequencies. It ought to be said that brain state cannot be defined rigorously without having specified a proper context which is endowed with both temporal and spatial scales in a particular experiment of theory [48].

2.2.1 The Search for the Meaningful Brain States

Neural correlates of mental states are points in the phase space of the nervous system that are associated with mental states. We can avoid the hard problem of neural properties and phenomenological states [49], if we focus on the identification and isolation of those points in the phase space, in order to discover under which conditions cognitive conditions arise and evolve in time [50]. One of the most promising strategies for the identification of neural correlates of cognition is the state space approach originating from the analysis of dynamical systems. It is still unclear how the variables that constitute the neural correlates of cognitive functions, such as perception, memory, language or consciousness, must be chosen. The study of the nature of this correlation is today a hot topic with a long and honorable history that goes back to Helmholtz and the first Freud.

The term neural code refers to the relationship between cognition and neural activity. Neural spikes are a mechanism for information transfer, of course other communication mechanisms exist at other levels, including the intracellular. Understanding the mechanisms and rules of neural coding is one of the most critical tasks in cognitive neuroscience. Yet, "neural code" is often used as a metaphor rather than as a well-defined object [51]. This is an indicator that researchers have not been able to provide a formal theory that makes unnecessary the use of these and other concepts in a metaphorical sense [52]. To be able to claim that we understand the neural code, we must show that we can reliably associate neural activity with a given situation and vice versa, that is, predict the current state of things on the basis of neural readings. Here the emphasis is on active neural coding, which cuts across the artificial divide of sensory and motor domains.

How does the neural code contribute to active perception, memory, and decision-making? The difficulties in characterizing the neural patterns associated with external and internal stimuli may be attributed to factors such as the high dimensionality and complexity in the neural space, and to technical limitations associated with measurement techniques. However, the dramatic improvement in neural recording and the extensive availability of computational resources, call for a reassessment of the neural coding problem. In order to understand how meaningful patterns are formed and disbanded in the brain, questions such as those cited below need to be addressed with a fresh eye. (1) How can the increase in the number of neurons simultaneously recorded improve the study of neural coding? (2) How do the current methods of neural data analysis, focusing on receptive fields, tuning curves or state spaces, need to be changed in order to gain a new understanding of neural coding? (3) Are spikes the only relevant information source to study neural code? (4) How can a new approach to the neural coding problem shed light on the neural mechanisms that mediate cognition? (5) How does the concept of active neural coding differ from the classic concept of representation? (6) How can behavioral analysis be combined with neural recording to explore the functions of the neural code?

2.3 Modeling Neurons

In the brain, the microscopic level is that of the neurons and its components. Neurons are not the most numerous cells in the brain, for example there are more glial cells than neurons, but only neurons can transmit electrical signal over long distances. Hence, to talk about how the brain processes information is to talk about neurons exchanging information. It is important, and maybe illuminating for some scholars, to realize that a neuron is a complex dynamical system that can be studied in terms of very different aspects. For example, while neurobiologists are interested in measuring the membrane potential conditional on voltage-gated ion channels, pharmacologists deal with the chemical blockers of those channels, and computer scientists may want to model neurons as digital gates [53].

Focusing now in mathematical modeling of neural dynamics and for the sake of clarity, I will make a distinction between two general approaches for the construction of models of neurons. On the one hand, we have biophysical models of neurons, the Hodgkin-Huxley and the Rall models are well known examples [54,55]. On the other hand, there are more simple models of neurons with a reduced parametric space. The "integrate-and-fire" model (I-F) is an example of this type of "simple" model. The I-F model is based on the concept of action potential as a threshold process. In this view, the spatial dimension of the neuron is not taken into account, neurons are seen as unit points, and only temporal aspects of the input-output properties are included in the model. Let us see some examples of these two kinds of models of neurons.

2.3.1 Detailed Biophysical Models of Neurons

In 1952 Hodgkin and Huxley, experimenting on the giant axon of the squid, found the ionic mechanisms underlying the initiation and propagation of action potentials. They discovered different types of ion-currents in the nerve cells: sodium Na^+, potassium K^+ and chlorine Cl^-. The ion channels of Na^+ and K^+ control the flow of those ions through the cell membrane, and mediate in the creation of action potentials which are electrical pulses that allow the transmission of information within nerve cells. Formally, the Hodgkin-Huxley model consists of a equation for the total membrane current, based on the assumption that the cell membrane has an intrinsic capacitance. The Hodgkin-Huxley model aims to put in mathematical terms the major features of membrane nonlinearity i.e., the voltage-dependent ionic currents that activate and inactivate in time.

$$I_M = C_m \frac{dV}{dt} + I_K + I_{Na} + I_L \tag{2.1}$$

The Hodgkin-Huxley equation is based on the Ohm's law, the left side of Eq. 2.1 is the total membrane current, I_M, and the right side includes the capacitance of the membrane C_m, the membrane voltage V, the sodium current I_{Na}, the potassium current, I_K and I_L, which is the leakage current carried by other ions that move passively through the membrane. The relevance of this model is unquestionable. Indeed, the Hodgkin-Huxley model is usually taken as the origin of computational neuroscience [56], for Denis Noble it is even at the genesis of systems biology [34].

More detailed models came with the introduction of cable theory in neuroscience. This theory is originally due to Lord Kelvin's theoretical analysis of signal attenuation in telephone cable. The phone cable is made of copper, which is a much better conductor than the salt solution inside a neuron. However, cable theory has been successfully applied for quantitative predictions about the attenuation of post-synaptic potentials (PSP) as they propagate through a length of dendrite towards the soma [57]. Wilfred Rall, Denis Noble, Christof Koch and others have applied cable theory to nerve fibers. Neuronal cable theory refers to the set of assumptions and results relating to the propagation and interaction of electrical signals in spatially extended nerve cells [58]. The work of Rall is particularly relevant because he demonstrated, using cable properties, that an usual assumption in neural modeling i.e., neurons are isopotential, was flawed. Rall was the first one to take into account the current flow to dendrites in the models of neurons, before him, the electrophysiological importance of dendrites was ignored [55, 59]. For a quantitative study of the role played by dendrites in neural function see [60] and [61].

Stochastic neuron analysis is being used to model noisy neurons i.e., noise involved in synaptic transmission. Green's functions, a type of function used to solve inhomogeneous differential equations, could be a possible solution for the cable equation because it allows to take account time delays in neuronal transmissions due to the non-trivial geometry of most neurons in the brain. Further developments in cable theory include temporal modifications in neuron transmission due to non-trivial neuronal geometry (specifically the branching dendrites) [62, 63].

2.3.2 Models of Neurons Based on Threshold

Integrate and fire models (I-F) are essentially threshold models of neurons. Already in 1907, Lapicque [64] stated that action potentials are generated when the integrated synaptic inputs to a neuron reach a threshold value. The integrate and fire models are based on this idea. I-F models provide a phenomenological description of neural behavior, but they are not explanatory in the sense that they do not clarify the underlying biophysical causes of electrical activity. The leaky integrate-and-fire (LIF) model [65] is a variant of the integrate and fire models. In a LIF approach it is assumed that the non-stationary temporal evolution of the spiking dynamics can be captured by point-like models of neurons [66]. The problem with this kind of model is twofold, one it depends on the value of the threshold which is rather arbitrary, and two, it includes a reset after each spike generation which makes the whole dynamics highly nonlinear.

The I-F models, compared to the biophysical models discussed in Sect. 2.3.1, have less parameters, which make them inadequate to predict, for example, how the neuron's behavior is affected by changes in the extracellular temperature or how it reacts to pharmacological manipulations [67]. While I-F models seem to be more efficient for investigating collective phenomena in neuronal networks [68], the biophysical models are more accurate in describing real single neurons [69]. However, the Achilles' heels of biophysical models is that we still lack a systemic methodology for measuring a large number of parameters. The blue brain project is working in that direction [70]. It is expected that in the next years, technological innovations in imaging techniques will open new possibilities in quantitative biophysical models of nerve cells and systems. The systematic combination of techniques such as infrared microscopy and multineuron patch-clamping is dramatically fostering our understanding of the molecular, structural and electrical properties of neurons and their synapses [71].

2.4 Modeling Populations of Neurons

To claim that neurons are the computational units in brain circuits and that the representational power of the brain is dependent on the microscopic level, is an uncontroversial minor statement. In order to give responses to questions like how the brain perceives the meanings of the stimuli or how the categories of perceptions are built, we need to go further. We must differentiate between the ability to sense the outside world – the problem of recognition – from that of building a representation based on what the surface of brain cells sense – the problem of representation. This second problem is what allows us to translate the information in the world into something meaningful for the brain. We do not fully understand the second problem, but we are beginning to see what some of the rules are. For example, they involve the coordinated effort of a largely parallel and distributed architecture of neural assemblies [72].

This section addresses the mesoscopic and macroscopic levels of brain modeling, the former is in between the microscopic level of single neurons, and the macroscopic level which deals with the global understanding of dynamics by the study of whole brain areas and hemispheres. While to measure the electric activity of a single nerve cell e.g., patch-clamp technique, needs a much more complex experimental setting than measuring the average activity of large cell ensembles e.g., EEG, to understand the nature of a specific neuron signal within a neuron assembly is a formidable challenge. To measure assemblies of neurons, we use techniques like EEG or MEG which record the electric and magnetic activity of hundreds of thousands of units. For example, an electrode placed on the scalp of an individual, records a summation of myriad of electric events in the neighborhood of the electrode. Thus, in order to deal with the formidable number of different neural events involved in the recordings, we need to have available mathematical structures to cope with the aggregative nature of the signal.

Let us now start presenting the different approaches for the study of the dynamics of interacting ensembles of neurons. Complex dynamic function cannot be understood by observing the parts in isolation. Integrative functions emerge from the dynamics of entire networks, rather than from linear computations performed at each nodal point of the circuit. To understand cognition we need to understand large scale neural networks.

2.4.1 Artificial Neural Networks

One of the most prolific mesoscopic approaches is the Artificial Neural Network (ANN), originally created by the neurophysiologist Warren McCulloch and the mathematician Walter Pitts [73]. McCulloch and Pitts, conceived the digital computer as a suitable medium for implementing brain processes. In their view, both the computer and the brain shared a binary nature, the nervous system operations could be reduced to a set of binary states, 0 and 1. The state 1 corresponds to a firing neuron and 0 state to a non firing one. In cognitive neuropsychology the ANN approach is concomitant with connectionism, which is usually presented as a reaction to the "failure" of the symbolist approach in Artificial Intelligence. The bibliography of the symbolist-connectionist clash is wide, and for the most part poses stimulating philosophical problems, such as the Chinese room argument [74], the symbol grounding problem [75] or the Turing test [76]. Connectionist models use linear algebra applied to point-neurons. It might be remarked here that linear systems analysis is arguably one of the major achievements of twentieth century mathematics. However, it is questionable that linear analysis is a suitable strategy in brain science. For example, in EEG data analysis, complex phenomena such as the self sustaining activity of the brain cannot be entirely explained in linear analysis basis.

In [77], the interrogation: are artificial neural networks (ANN) a good model for cognition? is answered negatively. William C. Hoffman, a mathematician, identifies the weakness of the connectionist approach in the following points: (i) the omission of the topological structure fundamental for the information processing, (ii) omission of the morphological structure of neurons, (iii) the inability to perform memory retrieval from point-neuron "weights" in neurobiological real-time, and (iv) the failure to implement psychological constancy. In Hoffman's view, these limitations taken together, disallow connectionism as a valid approach for a theory of cognition. Some philosophers of mind have also take a negative stance on connectionism. For example, in Fodor's opinion the neural network approach is ill-suited for a theory of cognition. Fodor points out that connectionist models cannot handle compositional semantics and as this is a property that minds have, it follows that connectionism is disqualified as a valid theory of mind [78]. Aizawa goes a bit further to claim that connectionism in cognitive psychology ought to be abandoned because connectionism is committed with the wrong level of analyses for cognition [79].

Despite the criticisms, the importance of ANN is beyond question, whole research fields like cognitive neuroscience would not exist without ANN. Furthermore, the range of applications of ANN is massive: decision making, pattern recognition, medical diagnoses, speech recognition or robot modeling and control, just to cite a few. The rationale of this must be found in one capital fact: ANN can implement any nonlinear dynamical system [80].

Non-parametric estimators can be arbitrarily well approximated, this property is called consistency and it is the major reason of the popularity of non-parametric estimators such as neural networks or Boltzmann machines. Consistency guarantees that, for a sufficiently big training data set, non-parametric estimators achieve the best possible performance for any learning task. It is important to note that the condition "sufficiently big training data set" implies that consistency is an asymptotic property, which can be formally stated as follows:

$$\lim_{n \to \infty} E[f(x) - E[y|x]] = 0 \tag{2.2}$$

In words, Eq. 2.2 non-parametric estimators $f(x)$, are consistent for all regression problems $E[y|x]$ (the mean value of Y given X). Thus the function $f(x_i)$ must minimize the loss function $\sum_{i=1}^{n}(y_i - f(x_i))^2$, if for example, we use the least square method. But there is a toll to pay here. The versatility of the estimator to optimally approximate any task is necessarily sensitive to the characteristics of the data. For example, when data samples are small or have dispersed distribution, parametric estimators may outperform non-parametric ones. Indeed, non-parametric estimators are optimal because they are consistent, but consistency is an asymptotic property.

2.4.2 Neurodynamics

Neurodynamics is the study of neural oscillations using differential equations. Neurodynamics builds feedback models that cope with lower levels activity that self-organise into higher level ones, forming mesoscopic states that constraint the phase space by reducing the number of degrees of freedom from the original microscopic configuration. This area of research is strongly interdisciplinary, as it shares tools and insights from fields such as theoretical neurobiology, nonlinear dynamics and statistical physics.

The mechanisms by which changes in the electric membrane potential trigger action potentials in neurons are well known since Hodgkin-Huxley model, which was described in Sect. 2.3.1. A sequence of action potentials generated by neurons is called spike train. The oscillatory patterns of spike trains that arise in brain activity can be studied using mathematical tools such as harmonic oscillators. Arguably, periodic spiking is at the basis of neural coding. However it might be mentioned that scientists have detected subthreshold membrane potential oscillations that deserve to be studied because they may also contribute to oscillatory activity by facilitating synchronous activity of neighboring neurons (Fig. 2.2).

The idea that neurons pulses are point processes is still prevailing. In this light, neurons are conceived as integrators with a threshold. If the sum of the incoming spikes from other neurons is above the threshold, the neuron fires, otherwise the neuron remains quiescent. The digital (discrete) bias in this conceptualisation of the dynamics of neurons is unmistakable. Neurodynamics represents an alternative to the oversimplified view of neurons as digital gates – a neuron is at *rest* or *firing* – in the connectionist tradition. Neurons are dynamical systems, this conception was clear in the original model of Hodggin-Huexley, and it is a key feature that might not be overseen. Neurodynamics makes extensive use of dynamical systems theory to extend the repertoire of mathematical tools in neural modeling. For example, the *resting state* would correspond to *stable of equilibrium* and the spiking state can be explained as a *limit cycle attractor*. Furthermore, the probability that a neuron fires increases if the equilibrium state is near to a *bifurcation* [81].

In neurodynamics, the final objective is no other than the construction of dynamic models of brain activity, but why do we need to coin a new discipline when we already have classical dynamics which has a long tradition of dynamic modeling? In classical physics, initial conditions and external forces are controlled by the experimenter, and the system can be modeled using feedforward differential equations that embodies the linear causality between inputs and outputs in the system. For example, the Newton-Euler equations of an open kinematic chain shown in Fig. 2.3. Neurodynamics, on the other hand, deals with brains which cannot be modeled in this way because causality is not always directly proportional.

Traditionally, in task-related studies the resting state is assumed to be the baseline or control state with no functional significance. However, this approach is not complete because it does not take into account the endogenous activity of the brain

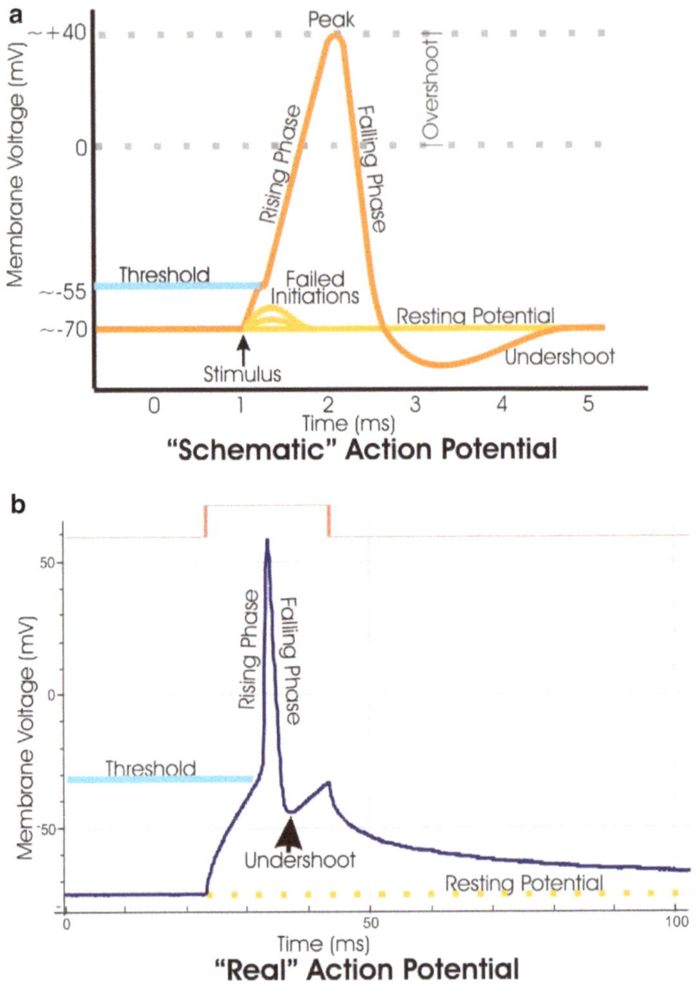

Fig. 2.2 Amplitude and duration – expressed in mV and ms respectively – determine the action potential of a nerve cell (**a**). The figure shows an idealized action potential in (**b**). The recordings of action potentials are often distorted compared to the schematic view because of variations in electrophysiological recording techniques. The membrane potential under the threshold may also affect to oscillatory activity of neighboring neurons (Image borrowed from [11])

which is active even in the absence of explicit stimuli. The classical analysis based on task-evoked responses, may lead to wrongly conceive the brain as a reflexive system. This is at odds with a simple fact, the difference between the brain's energy consumption in the resting state is often only a 5 % less than when it is engaged in some attention-demanding task [82]. This may lead us to rethink the concept of

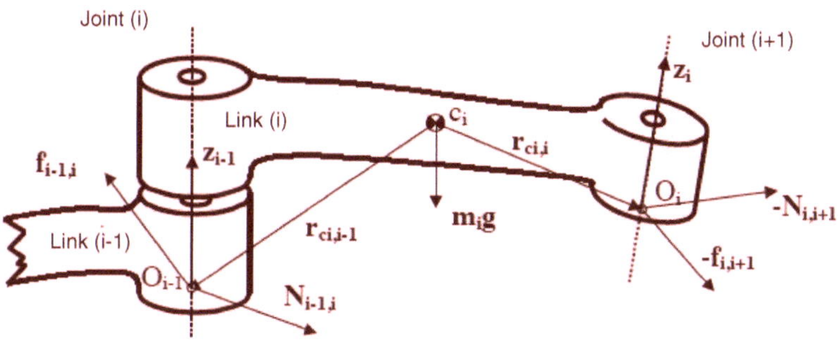

Fig. 2.3 The figures depicts the free body diagram of Link (i), that is part of a manipulator arm with a number n of Links, where each two Links (i) and (i-1) are connected through a Joint (i). The Newton-Euler Dynamics model allows to find the force and torque required to move the robot following a specific trajectory. The force $f_{i-1,i}$ and moment $N_{i-1,i}$ that Link (i-1) applies to Link (i) at Join (i) are iteratively computed. This methodology is characterized by transcendental, non-linear systems of differential equations that can be solved numerically. The analytical solution can be extremely challenging

brain's resting state. There is enough evidence that the human brain is endowed with a rich dynamics even in the absence of external stimuli. For example, recent advances in functional magnetic resonance imaging (fMRI) techniques have showed that the baseline state of the brain is by no means an inactive state [83, 84]. Slow (<0.1 Hz), spontaneous fluctuations of the blood oxygen level dependent (BOLD) signal in fMRI are highly coherent within either structural or functional networks in the human brain. Therefore, exploring these fluctuations could lead to a better understanding of the brain's intrinsic or spontaneous neural activity. Thus, we must distinguish between induced activity which is triggered by external stimuli or motor responses and ongoing or spontaneous activity which is the brain activity that cannot be associated with external stimuli or specific events. Neurodynamics is capable of incorporating the spontaneous activity in its models, and relate it to mental states such as sleep [85], rather than label it as noise which is commonly done in studies focused on stimulus-related activity.

Statistical mechanics or statistical physics has been increasingly extending its range of applicability outside the domain in which was created, to deal with applications as disparate as metabolic scaling [86] or the analysis of social nets [87]. A word of caution is in order here, statistical mechanics applies to classical systems close to equilibrium, but brains are dissipative and far from equilibrium systems. It ought to be remarked that the brain is a dissipative system. Dissipative systems have bounded free supply (energy, matter), that is to say, the amount of energy or matter that flows out of the system to the environment cannot be arbitrarily large [88]. From a thermodynamic setting, all systems are dissipative because subjected to the second law of thermodynamics. Thus, dissipativity is a property of open systems such as the brain.

2.4.3 Neural Field Models and Neural Masses

Neural field models use mean-field approximation, describing the activity of large number of neurons by building models with "simplifying assumptions and empirical priors [89]". The mean-field approach enables to sidestep the problem of modeling systems with very high dimensional space. The basis of this kind of models rests on the probability density over ensembles of neurons. The Fokker-Planck equation is used to calculate the density dynamics. Thus, density-dynamics and mean-field approximation are the basis of neural-field models. It might be noted that while this kind of models have shown to be well equipped to describe the steady-state regime in a system, their appropriateness in non stationary behaviors is under debate [90]. Neural field models are a generalisation of neural mass models, or put in a different way, the latter are a particular case of neural field models. As we will see below, the Fokker-Planck equation allows to model populations of neurons using the probability density: when in a given model, the density represents a point-mass over the space of states of a population of neurons, the model is called neural-mass model. On the other hand, when the location of the mass or expected state of the ensemble of neurons is not a point, but a function of both time and position on the brain, the model is referred as neural field. In the rest of the section the basis of neural field model and the Fokker-Planck equation are introduced. Moreover, references of major results in neural mass models are provided.

Neural field models borrow concepts from statistical mechanics. The main idea behind this approach is to reduce the number of the degrees of freedom in a population of neurons i.e., reduce the state space dimensionality. In this context, a degree of freedom is a parameter (a variable) measured for each neuron. This is a well known procedure, for example, the Boltzmann's theory of gases, gets to reduce the individual degrees of freedom of the single particles by means of the introduction of field terms, like temperature or pressure. In a similar way, a large population of spiking neurons can be reduced to a function of distribution that captures the neural state of the whole population for any given instant. Thus, for a population of neurons of any size, we can calculate the density of the phase state $\phi(V, R)$, where V is the potential and R the membrane's resistance. The Fokker-Planck equation gives the density dynamics of large populations of either molecules in a gas or neurons in a brain area. The equation, irrespective of the dynamics of the particular components, is linear and quite simple. What is gained with this method is that even though the dynamics of a single neuron may be highly complex, the density is simple, linear and deterministic. It is worth mentioning that chaos theory is not a stable theory as can be quantum theory or electromagnetic theory, but it provides a set of mathematical tools that may be used to model non linear phenomena.

In the neural context, the Fokker-Planck equation describes the time evolution of the probability density function of the state of a neuron. For example, if we define the state of a neuron as $q = (V, I, t)$, where V is the membrane voltage, I the current and t the lapsed time since the last action potential, then the density of the phase space defined by the ensemble of neurons is $p(Q, t)$, where $p : \mathbb{R}^n \times \mathbb{R}_+ \to \mathbb{R}_{0,1}$, is a scalar function that returns the probability density at each point in the phase space. The Fokker-Planck equation allows going from models of individual spike neurons (points in the phase state) into electrical waves (fields), that convey the spikes of myriad of neurons. Thus, it is possible to describe neuronal population dynamics in terms of the Fokker-Planck equation. As an example of this approach, Breakspear and Stam [91] have constructed a neural dynamics model based on the Fokker-Planck equation that tackle the multiscale hierarchy of levels in the brain.

Neural mass models capture the dynamics of a neural population. The term was used by W.J. Freeman in his 1975 book "Mass action in the nervous system [92]". The theory of K-sets developed by Freeman in the 1970s, is based on a nested hierarchy of models that describe the dynamic behavior of neural masses at the mesoscopic level. A neural mass refers to a neural system around 10^4 neurons with 10^8 synapses. These models captured the dynamics of neuronal ensembles in the olfactory bulb [93]. Neural mass models are generalized in [66] by furnishing wave equations that describe the spatio-temporal evolution of the expected neural states which are here a function of space. A common feature of the models described in this section is that the critical step is to identify the variables at the mesoscopic level to then write the equations that express how the activity of self-organised neural ensembles produce global macroscopic order. A new paradigm is emerging, providing a new understanding in terms of transient activity that connect different brain states or attractors, rather than in the classical approach of linear dynamics, that expects that brain activity will be confined in an attractor as a response to controlled perturbation induced from the outside. Thus, the focus is put on the passage among attractors, rather than on the study of equilibrium states that act as sinks where the system's activity is confined. This opens a new window of comprehension of complex systems and in particular the brain.

2.4.4 Biological Synchronization: The Theory of Coupled Oscillators

In this section, I outline the theory of coupled oscillators focusing on how this theory deals with coupled dynamical systems such as neural assemblies. Phase interactions between oscillators characterise a big variety of phenomena, including biological ones. It ought to be noted that synchronicity is a particular case of phase locking. We say that when two oscillators are synchronized the influence between them is minimal [94]. Prior to understand coupled oscillators, it is necessary to specify what an oscillator is and how it works. An oscillator is any system with a periodic behavior, that is to say, a system that after a period T, returns to the original state, $f(X,t) = f(X,t+T)$.

The theory of coupled oscillators has been applied to brain coordination dynamics [95, 96]. The coordination is empirically observable in terms of the synchronisation of the neural firing in neural populations. In [97] is conjectured that the symmetry in the gaits of animal locomotion e.g., walk, trot, pace, etc., suggest a

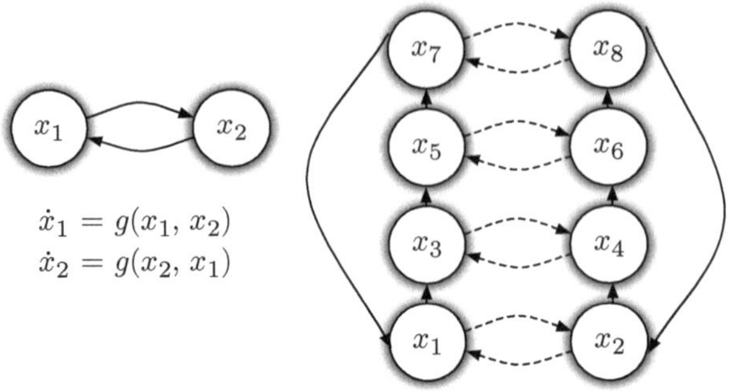

Fig. 2.4 The simplest coupled cell network is the two-cell system depicted on the *left side* of the figure. The class of differential equations associated to this network is $\dot{x_1} = g(x_1, x_2), \dot{x_2} = g(x_2, x_1)$ where $x_1, x_2 \in \Re^k$ are the state variables of the cells. Note that the system is as simple as to be defined with one single function g. Symmetry in animal gait shows that for quadruped animals, the only possible symmetry class of CPG networks is the 8-cell network shown on the *right side* of the figure (Following [12])

structure for locomotor central pattern generators (CPG). This hypothesis states that a population of neurons governs the rhythms associated to the various locomotive gaits. For an overview of models that exploit the symmetrical characteristic of the animal gaits, see [98, 99]. It ought be remarked that the work on central pattern generators is primarily theoretical. The identification of ensembles of neurons that orchestrate the rhythms involved with gaits or other periodic activities in the body of systems far more complex than a leech, is more of a mathematician's dream than a biological reality.

In [12], Golubitsky discusses the importance of symmetry in neuroscience. This a very important work from a theoretical-mathematical perspective, because it gives a mathematical description of the symmetries found in the different phenomena relevant for natural scientists: animal locomotion, the connectivity of visual cortex and the vestibular system are described. The use of symmetry in neuroscience is a very promising direction. For example, interesting examples of this line of research are [100], where the symmetry in the leech's heart is tackled with ordinary differential equations and Grupoid theory. Theunissen et al. have developed a mathematical formalism on the neural coding in the cricket cercal sensory system [101].

A network of oscillators is described as a graph, whose nodes represent neurons and whose arrows indicate which neurons are coupled to which. Each neuron is a system of ordinary differential equations (Fig. 2.4). The theory of symmetrical Hopf bifurcation [102] gives the means to classify the patterns for network oscillators. The application of the theory of coupled oscillators in brain function studies relies upon the assumption that nerve cells are intrinsic oscillators, weakly and globally coupled.

Synchronisation is a mechanism for the transmission of information between assemblies of cells with converging connections. For example, the synfire hypothesis is based on this idea [103]. Synchrony in the context of neural firing means that there is a synchronous occurrence of spikes for a given period or phase. The functional role played by synchronisation in coding information is under debate and an unified theory of neural coding has not been attained yet. There are two main hypothesis in neural coding. The rate coding hypothesis assumes that the information about the stimulus is conveyed in the mean firing rate. The temporal coding hypothesis, on the other hand, argues that behaviorally relevant information resides in the precise time of firing, which is necessarily missing in the rate code approach as it averages the spikes. The first assumption – neurons are intrinsic oscillators – deserves attention. The existence of intrinsic oscillators has been reported in neurobiology, some notable examples are the neurons in the inferior olive nucleus or the neurons in the nucleus reticularis in the thalamus [104], which are able to generate a rhythmic activity with a wave frequency of 8–14 Hz [105]. Caplan et al. [106] have shown that theta oscillations (4–7 Hz) in neurons in the hippocampus are involved in spatial learning and navigation. However, in the nervous system, these are the exception more than the norm. Major characteristics of brain dynamics like transience, variability and non-stationarity in neural firing, represent a limitation to oscillator theory in the brain. While neurons, with the exception of pacemakers, are not exactly intrinsic oscillators, neuronal networks have a propensity to synchronize. Synchronization is a robust feature present in normal function but also in brain pathologies e.g., epileptic seizures show enhanced local synchrony [107]. Thus, oscillatory models represent a valid but incomplete approach to understand and characterize the synchronous emergent behavior of neural assemblies. The second assumption – neurons are weakly coupled – is uncontroversial. As everyone knows, neurons are chemically and electrically coupled by neurotransmitters and synapses. The third requirement – neurons are globally coupled – is not, broadly speaking, a realistic assumption. Glial cells, on the other hand, show global coupling via gap junctions [108]. However, oscillations are still real in brain activity. Neurons have different firing patterns, depending on the cognitive state. The different behavioural and perceptual states of the brain are associated with different brain rhythms. Resonance is a property that describes the ability of neurons to respond selectively to inputs at preferred frequencies. For a general study of the relationship between electrical oscillations and resonance in neurons, see [109]. In [110] a more detailed case of intrinsic resonance frequencies is shown in the stellate cells in layer II of medial enthorinal cortex. It may be remarked that in less controlled scenarios, the noise and rapid fluctuations in brain activity makes the search for the intrinsic oscillators in the brain extremely complicated.

The seminal works of Winfree [111] and Kuramoto [112] set the basis for the study of synchronization in coupled oscillatory biological systems. In essence, the models are based on the reduction of the cell's activity to a phase i.e., θ_i within a cell network, $\theta_{1...n}$. It might be remarked that prior to derive the phase or the amplitude in the signal, we need first to observe a periodic activity which is sometimes challenging given that variability, non stationarity and noise are prevalent features in the electrical recording of the brain. The Kuramoto model captures the time evolution of the phase of the oscillator in the formula 2.3

$$\frac{d\theta_i}{dt} = \omega_i + \sum C_{ij}(\theta) \tag{2.3}$$

where ω_i is the intrinsic frequency and C is the coupling function which needs to be calculated using, for example, averaging methods [113]. The Haken-Kelso-Benz, described in Sect. 2.5.2, is another example of a model based on low dimensional description of systems of coupled differential equations in terms of a few parameters or order parameters in the Haken's notation.

In conclusion, the requirements of coupled oscillatory mechanisms may not be clearly observed in neurons tissue but network synchronization is a robust feature of brain function which makes the oscillatory framework a valid approach to tackle this emergent phenomenon [114]. However, as it has been pointed out by Buzsaki [115], whether two or more events are synchronized or not, can only be determined by their impact on the pertinent reader mechanisms able to interpret the signal that is been transmitted, as for example, in synchronous neural firing. Accordingly, the goal of synchronous spikes can only be understood when there is a reader mechanism – a neuron or group of neurons – that is able to integrate the incoming firing events over a time period that is specified in the membrane time constant.

2.5 Large-Scale Brain Modeling

A common feature of the models described in Sect. 2.4.2, is that the critical step is to identify the mesoscopic state variables that refer to local masses of neurons. In this section, the focus is on the spatio-temporal patterns of brain activity at the macroscopic level. I will sketch some mathematical tools that promise to leverage our understanding of the brain at a system level. In order to cope effectively with the brain emergent properties that mediate in high cognitive functions, it is necessary to have an understanding of the interactions between different brain regions. The dynamics of the brain is high dimensional and non linear, therefore we need mathematical tools able to cope with the extreme complexity of brain organisation. Let us start now by sketching some mathematical tools that are tackling the problem of brain dynamics at mesoscopic and macroscopic scales.

2.5.1 Theory of Attractors in Brain Dynamics

The theory of coupled oscillators that was sketched in Sect. 2.4.4 is a theoretical perspective that, so far, cannot deal with the empirical evidence that shows that the rich internal brain dynamics is characterized by stochastic and chaotic properties.

The internal brain dynamics cannot be entirely explained within the linear analysis paradigm, the rationale for this limitation is two-fold. First, the linear-stimulus response paradigm is powerless to deal with mental phenomena such as phantom limbs or hallucinations, which are processes that arise from internal brain state rather than triggered from external stimuli. Second, brain activity is characterised by abrupt state transitions. In EEG and MEG activity, fast and robust sensory processing have been observed through frequent phase transitions [116]. Ordinary differential equations have the serious limitation that they cannot describe phase transitions, they are good at describing what happens in the system before and after but not across the transition. A new approach that emphasizes the transient activity that connect different brain states or attractors, rather than the study of equilibrium states – attractors where brain's activity is confined in response to controlled perturbation induced from the outside – is required. Furthermore, the dynamics of complex vertebrate systems, in contrast to more primitive systems characterized by very stable patterns of coordinated activity [117], are governed by stochastic dynamics and chaotic regimes [118–120]. For a critical discussion about the biological relevance of theoretical dynamical regimes in brain dynamics see [121].

An attractor is a set towards which a dynamical system evolves over time. An attractor can have several geometric forms such as point, plane, ... or even a recurrent fractal structure, in which case it is called strange attractor. To define the brain as a dynamical system without describing its attractors is like building a tree house without a tree. The study of attractors in neuroscience is rooted in the idea that stable, persistent activity is a key aspect to understand neural computation. The formal definition of an attractor as a closed subset lacks a straightforward translation in terms of physiological recordings time series. As a consequence of this vagueness in the interpretation, the term attractor has been used in a rather loose way. For example, in [121] it is argued that in neuroscience, we should speak of attractor-like systems rather than attractors.

In [122], temporal correlations between stimuli are converted into spatial correlations between attractors. Cossart et al. [123] have investigated the spatio-temporal dynamics of neocortical activity, concluding that co-ordinated activity of groups of neurons may correspond to circuit attractors-emergent features of feedback neural networks. The most spectacular finding related to attractors is cells that display persistent activity in memory tasks. As early as 1949, with Donald Hebb's inescapable book "The organization of behavior" [124], short-memories have been thought to be attractor states of neuronal representations. In this view, memories could be modeled as fixed point attractors. More recently, attractor dynamics has been used to explain context-dependent episodic memories in the hippocampus [125].

It is important to realize that the empirical demonstration of an attractor is, roughly speaking, what engineers want to know when looking at how a system works: perturb the system in order to study how it comes back to its resting state. EEG activity is enormously robust, for example, epileptic seizures are hard to destabilise once they have started. Thus, it must be seen as logical that the theory of attractors is being used to explain the dynamics of brain disorders such as epilepsy

[126, 127] or Parkinson's disease [128]. However, there is an important caveat. The view that brain activity is characterised by abrupt state transitions conveys a preferential epistemic status to transient regime versus invariant attractors in brain activity. Although the search for invariant attractors have not succeed, the reason for this failure may be found in the insufficiency of the theoretical/methodological available means to address the *invariance problem* in brain activity, especially at the higher levels. Thus, the crux of the problem may be methodological (are we using the good tools to detect invariance in brain activity?) rather than ontological (do invariant attractors exist?) [121]. A possible way out to this conundrum is to focus on simple systems like invertebrate nervous systems in which properties such as robustness and synchronous coupling, can be more efficiently addressed [129, 130].

2.5.2 Synergetics

While the whole brain dynamics is high-dimensional, low-dimensional dynamics coexist in specific brain areas, for example the inhibitory interneurons of the hippocampus cells, also called *theta cells* because their firing is according to the ongoing theta rhythm[1] [131]. This has lead some researchers to hypothesize that brain dynamics is chaotic [132, 133].

Hermann Haken, a laser physicist, is the creator of the interdisciplinary field of research called synergetics [134, 135]. Synergetics is the study of the general principles of self-organization in systems, irrespective of the nature of their individual parts. In Haken's synergetics, causality has two directions, downwards and upwards. The upwards causality is local-to-global – novel systemic dynamics emerge from the interaction of close components. The original aspect of synergetics comes from the downwards causality which is global-to-local. The slaving principle states that there are global order parameters that enslave the constituents of the system and effectively govern local interactions between them. This principle may help to explain the low-dimensionality of neural ensembles in the brain. Because the cooperation between the individual parts, one speaks of circular causality. At a critical point, a single order parameter may undergo a non-equilibrium phase transition (see bifurcation) with symmetry breaking, critical slowing down and critical fluctuations.

[1] Theta oscillation can be used with two different interpretations. In rats and mice, theta oscillation is a specific type of regular oscillation seen in the hippocampus and several other brain regions connected to it. Thus in rats we say "hippocampal theta oscillation". The original sense of theta is used in human experimentation meaning: EEG waves falling into a frequency range of 4–7 Hz, regardless of where in the brain they occur or what their functional significance. Therefore "theta oscillation" and "hippocampal theta rhythm" are different because in the first case the EEG waves are obtained through electrodes glued to the scalp measuring several areas of the cortex, while in the latter case is implicitly assumed that the EEG wave comes from the hippocampus, using invasive techniques in rodents.

In synergetics, co-operation rather than competition, is hypothesised as the key mechanism that enables the existence of order parameters that determine the behaviour of the individual parts of the system. The order parameters are collective or systemic variables that arise when the system is in the brink of instability. The order parameters obey low dimensional dynamics and characterize the system macroscopically. In neuroscience, synergetics and in particular the slaving principle is being applied to study brain function. In this approach, the brain is conceived as a self-organizing system, operating close to instabilities governed by order parameters that enslave the individual parts or neurons [136]. The Haken-Kelso-Benz model for interlimb coordination e.g., finger taping has successfully study the stability of motion patterns using the relative phase of rhythmic finger movements as an order parameter [137]. This model has been refined by Frank in a stochastic framework using the Fokker-Planck equation. The recorded electrical activity in the cortex can be reproduced with stationary solutions of the Fokker-Planck equation [138]. However, the derivation of the phase equation poses serious challenges, specially when we want to model phenomena other than those related to simple motor actions. Furthermore, essential features in brain recording such as noise or non stationarity may need to be addressed with a new perspective that, for example, reconsiders the role played by noise in information processing i.e., noise is not necessarily a nuisance to be eliminated but it may convey valuable information [139].

2.5.3 Dynamic Geometry

The tensor network theory, proposed by Pellionisz and Llinás [140], proposes a geometric interpretation of brain function. The relation between the brain and the external world is defined in terms of an interactive geometrical relationship between sensory and motor expression. This theory relies on the idea that the brain's internal functional space must be able to interact successfully with the "external reality". The internal space is conceptualized using a dynamic geometry approach, in which an event in the external world and the corresponding event in the internal space, are described as a geometric map. Prior to defining any metric tensor one needs to have available a well-defined distance function. As the anatomy of the brain does not present a smooth and linear representation of the external world, we need to define a distance function which will be functionally isotropic i.e., the distance is the same in all directions.

The multidimensional spaces of the central neural system are, for the most part, not definable in well-known geometries such as Euclidean or Riemannian spaces. Thus, there must be a distance function defined using a stochastic metric able to deal with the stochastic nature of the functional space. The neurological reasoning behind this approach is as follows: local chaotic oscillations may give rise to synchronous oscillations on a larger scale [141, 142], therefore, from the local chaotic behavior and its associated stochastic metric may arise a smooth metric structure at the global level, for which there exists synchronous oscillations. Roy and Llinás have recently upgraded this theory [143]. The novelty is the use of a stochastic metric tensor, which is defined as five-dimensional space-time where the fifth dimension is a probability space. The stochastic metric tensor in dynamic geometry is defined as $g_{ij} = h_{ij}(x, \xi)$, where x is the four dimensional Minkowski space (3 for space + 1 for time), and ξ is a fifth dimension. The tensor is applied to discern between different dream states. In REM sleep $g_{ij} \neq 0$ because the internal degrees of freedom increase in the dream state. In deep sleep, the degrees of freedom seem to be frozen, $g_{ij} = 0$.

2.5.4 Network Theory

The study of complex networks focusing on network topologies and using graph theory concepts, has provided fresh insights into the problem of neural organisation. Since Watts and Strogatz' seminal article in Nature in 1998 about graph theory and its application to complex biological networks [144], biologists began to see network theory as a promising theoretical framework for quantifying the structural and functional properties of biological networks. Network theory provides a quantifiable framework, making explicit measures such as path length or node degree distribution to explore the structural and functional connection properties in the brain. It ought to be remarked that the existing network measures e.g., characteristic path length, modularity, betweenness centrality, motif z-score [145] etc. applied to brain connectivity might not necessarily be the most appropriate metrics for the analysis of brain networks, and new neurobiologically meaningful measures might need to be developed.

As it was pointed out in Sect. 2.5.1, methods based on differential equations have the serious limitation that they cannot describe phase transitions. The time-honour tradition for making simplifying model assumptions based on differential equation analyses, has been seriously challenged by the availability of fine grain representation of complex brain networks. Network theory has emerged as a transversal framework for data-driven model development of biological systems described in huge datasets. Differential equation models are unsuited to deal with a number of network phenomena – network modularity, the small world effect, transitivity, degree distributions, network resilience, efficiency in information processing related to graph invariants [146] – that are centrally important to understand the functioning and the malfunctioning – aging, disease – of the brain.

We can distinguish three main kind of network-based modeling in brain function: Artificial networks, Random and regular networks and Complex networks. *Artificial networks* refers to neural networks that were previously described in Sect. 2.4.1. These include purely feed-forward networks e.g., Perceptron [73], feed-forward back-propagation neural networks e.g., Hopfield's model [147] and related attractor networks [148]. *Random networks* are unstructured because any two nodes have the same probability to be connected independently of the distance. On the other hand, in a *regular network* each node has exactly the same number of links. The "structure" of a network can be quantified with the entropy function. Thus, the entropy of a random network is larger than the entropy of a regular network (lattice). Both regular and random networks are amenable to various types of analysis that borrows from statistical mechanics. For example, the Ising model – a mathematical model of ferromagnetism in statistical mechanics – has been used to model the activity of neurons in the brain. Each neuron is either active or inactive and arranged in a graph, usually a lattice, allowing each spin to interact with its neighbors. Using the principle of maximum entropy suggested by Jaynes [149,150], the average firing rate for each neuron in a given network can be calculated [151]. Special mention deserves discrete approaches to neural phase transitions. The phenomenology of phase transitions in neural processes are being developed using new mathematical tools. For example, Coupled Map Lattice (CML) [152, 153], Phase transitions in random graphs [154] and neuropercolation [155]. In particular, neuropercolation tries to generalize the cellular automata behavior to simulate the complex behavior of neural populations. Neuropercolation extends the concept of phase transitions to large interactive populations of nerve cells. The construction of a coherent framework that fuse the different approaches mentioned has not been achieved. However, empirical findings like neuronal avalanches [156, 157] or percolation transitions [158] may provide a solid experimental grounding to a researcher field with a strong theoretical outlook. Complex networks are different from random and regular networks which are mainly theoretical, in the sense that complex networks are data-driven realistic network models. Complex networks will be explored in detailed in Sect. 7.5. It is pertinent at his point to emphasize that the network-based approach is being pushed forward in different fields of biomedical research, however network biology is still in its infancy [159]. The hierarchical interdependency found in biological systems needs to be addressed taking into account the hierarchies and feedback loops among the network components. In order to acquire a system-wide understanding of the brain we need to be able to build models and theories that address the emergent properties that are conducive to its complex behavior.

2.6 Conclusions and Future Directions

The historically unique over abundance of data and the rich variability of mathematical models and computational tools available to the researcher in neural and cognitive science, must not led us astray from critical problems that are still in need of a solution. Arguably, the most important problems for the mathematics of brain science are those of going between levels to allow the manner in which overall behaviour arises from the fundamental level. However there is still the need to develop alternative descriptions of activity in the intermediate areas so as to make the bridging problems easier, as well as enabling more efficient testing of expected activity in these intermediate levels. New developments in such approaches as dynamical systems theory or stochastic dynamics are expected to give further power to their application to give a better understanding of brain processing. We must start thinking now, which answers or what kind of new questions will allow us to pose the complete reconstruction of the connectivity of human brain. The complete map of human brain connectivity will not convey in itself any mechanistic or probabilistic explanation of brain functioning. The problem of how cognition arises out of brain dynamics, can be studied as the formation and evolution of a special kind of patterns, those that are relevant for cognition. The understanding of those neural manifolds at multiple scales, will require the setup of a substantial mathematical and computational apparatus, maybe not devised yet. The mathematical theories relevant to the different scales has also been evolving, such as in the employment of category theory in providing new ways to model the emergence of cognition. This may open a new window of comprehension of complex systems and in particular the brain. In the successive chapters, Category theory is proposed as a new methodology that may allow to fuse the different existing frameworks into a coherent picture.

Chapter 3
The Categorical Imperative: Category Theory in Cognitive and Brain Science

3.1 Introduction

In this book is conjectured that category theory could provide the necessary concepts to bridge the gap between the different levels of brain activity: from microscopic neuronal activity, to macroscopic aspects, like for example, emotions or abstract thoughts, which are still imprecisely defined in the psychological literature. Traditionally, mathematical formalisms in cognitive science have been confined to toy model world descriptions [160, 161]. In the absence of a theory written in mathematical terms, the separation between the different disciplines that form the cognitive sciences, will be progressively more acute and an understanding between them unattainable.

It is unreasonable to think that purely linear mathematics can be the killer tool to model the complex interactions that take place in the brain. We may indeed, need a new mathematical language for describing brain activity. As Mac Lane, the creator together with Eilenberg of category theory, pointed out the fact that mathematics is protean, means that the same mathematical structure has many different empirical realizations, this is why mathematics is about patterns or forms, where each of these forms describes different aspects of the external world. Mathematics is therefore both the language and the conceptual toolkit to effectively deal with unlimited empirical contexts. This same relationship between mathematics and the external world, suggests a similar one between category theory and mathematics: all the mathematical fields can be organised according their structure by specific categories, and such specific categories can themselves be organised using the notion of general category as provided in the axioms of category theory. This chapter aims to set the agenda of category theory (CT) as the appropriate methodology that provides the necessary theoretical framework in order to build cognitive models in mathematical terms.

J. Gómez-Ramirez, *A New Foundation for Representation in Cognitive and Brain Science*, Springer Series in Cognitive and Neural Systems 7, DOI 10.1007/978-94-007-7738-5_3,

The structure and dynamics of the human brain are overwhelming – 10^{11} neurons with 10^4 connections each one. This gives an idea of the complications we are facing here. We can identify two major difficulties in modelling systems as complicated as the brain.

1. Undeniably the brain is a very malleable system, but the theories and models themselves do not have to be that malleable.
2. One of the biggest challenges in science today is to decipher the map of the brain. The view that the brain configures metric maps that reflect objects and processes from the world out there, needs to be revisited with a meaningful and formal concept of structure [162, 163].

Despite the daunting difficulties in understanding how cognition arises from neural interactions, it is possible to make advances. What is needed now more than ever in the brain sciences is to developed model-related analysis strategies that enable us to propose hypothesis that can be contrasted to empirical investigation. In the words of the mathematical biologist Michael C. Reed [164]:

> experimental neurobiology, like experimental physics, needs input from deep and imaginative theorists.

This chapter has three main objectives that will be explored as follows in the three following sections.

1. To introduce the mathematical theory of categories, providing formal definitions of key concepts (Sect. 3.2).
2. To outline the basis of a new theoretical framework based on CT that may cope with the binding problem (Sect. 3.3).
3. To provide a state of the art of the actual applications of this theory in the domain of cognitive and brain science (Sect. 3.4).

3.2 Category Theory

I will make in this section a historical sketch of the birth of category theory and the first categorists. In science, history tends to be dismissed as irrelevant. Nevertheless, the exclusive focus on the present and the last technical achievements is, I believe, nothing but detrimental. A historical perspective can strengthen our comprehension of the current concepts, improve our understanding of the problem, and help to foresee future developments.

Category theory is a new branch of mathematics i.e., abstract algebra, that was created by Mac Lane and Eilenberg (Fig. 3.1), as a way of studying and characterizing different kinds of mathematical structures in terms of their admissible transformations. The birth of category theory can be dated to 1945, with the paper by Eilenberg and Mac Lane entitled, "General theory of natural equivalences" [165], in which the notion of natural isomorphism is originally defined and used to capture

Fig. 3.1 Samuel Eilenberg, *left*, and Sanders Mac Lean, *right*, the two creators of category theory (Images from [13] and [14])

the shared structure in homology groups. By "natural" in natural isomorphism, Mac Lane and Eilenberg intended a common "phenomenon", for example a property existing in different contexts. In their original paper, group homomorphisms were "natural". Today it is common knowledge that the mathematical theory of categories deals with structures and the relationships between them. Each kind of structure – algebraic or topological – forms a mathematical category. Indeed, categories in mathematics arose of the need for the construction of a formalism to describe the passage from one type of mathematical structure to another [166].

It might be said that Mac Lane and Eilenberg underestimated the potential of their own creation, as they did not perceive category theory as "a field for further research effort, but just as a language of orientation" [167]. It was only later, and especially thanks to the efforts of Kan and Grothendieck (Fig. 3.2) that category theory was perceived as a radically different way of doing mathematics, far more important than the "handy language to be used by topologists and others", as Mac Lane initially intended. Gothendieck defined Abelian categories through axioms that, importantly, were not like the axioms for Abelian groups, but they constituted the description of Abelian groups and also of other similar categories. The rationale of Gothendieck's approach anticipated the current understanding of category theory in which the focus is shifted from the elements to the relationships between those elements. Thus, in category theory the nature of the objects are less important than the patterns described by the arrows that connect the objects.

Kan, in 1956, created the concept of adjoint functor which is now considered as the most significant contribution of category theory to mathematical thinking in its broad sense [168]. Kan's creation is a conceptual tool of the first magnitude on the par with the idea of continuous functions [169]. Ehresmann and Lawvere deserve special mention as well in the origins and development of category theory. Ehresmann, in the 1940s worked out the notion of local structures and later made major contributions in groupoids and structured categories. Lawvere's doctoral thesis in 1963 is a major milestone in the history of category theory. He proposed

Fig. 3.2 On the *left*, Alexander Grothendieck, a very remarkable man, Fields medal in 1966, disenchanted with world politics and academic circles, he moved into the woods, maybe somewhere in the Pyrenees. On the *right*, William Lawvere, the champion of category theory as a foundation for mathematics (Images from [15] and [16])

a new theoretical framework for mathematics itself, in which the set-theoretical framework is replaced by the axiomatic system provided by category theory. Lawvere's work is intrinsically related to the construction of a proper foundation for mathematics. Lawvere is a strong proponent of the theory of categories as the best possible foundation for the entire mathematical edifice.

Mac Lane followed with interest the work of Lawvere and revised his initial statement of category theory as a mere "handy language" to acknowledge a more relevant role for the theory he created. However, it must be said that Mac Lane never fully committed to a categorical foundation of mathematics as proposed by Lawvere, neither to the set-theoretic vision à la Bourbaki. In the beautiful book, "Mathematics Form and Function" [170], Mac Lane concludes that there is, as yet, no simple and adequate way of conceptually organizing all mathematics. Set theory, he points out, is a handy vehicle, but its constructions are sometimes artificial, while category theory, on the other hand, is almost everywhere in topology and algebra, but "do not as yet relate very well to most analysis".

Category theory can be seen as the "mathematics of mathematics" because one can make definitions in a broader context, that is to say, without reference to the particulars to which the definition might be applied. As we will see later, most constructions in mathematics may be defined by universal properties expressed in categorical terms. In contrast to the classical atomist approach of set theory that builds all its notions in terms of the elements, CT does it in terms of arrows. Indeed the motto of CT "thinks in arrows", means that CT is able to reformulate key concepts in terms of arrows or relationships rather than the elements themselves. However counterintuitive it may be, in category theory maps between objects rather than objects themselves take the primary role.

Category theory is inherently iterative: there are transformations between categories: namely functors, and transformations between functors called natural transformations. It is interesting to note that Mac Lane and Eilenberg originally defined "category" in order to be able to define "functor", and "functor" in order to be able to define "natural transformations". Thus, the theory of categories, functors and natural transformations, which is the actual name for Mac Lane and Eilenberg's theory, provides a characterization of the notion of structure-preserving transformation, and thereby of a species of structures admitting such transformations.

There are some excellent textbooks in CT. "Categories for the working Mathematician" by Mac Lane [171] is the "bible" of category theory, a compact and elegant textbook, but not easily accessible for readers not familiarised with category theory. More basic introductory texts are: "Conceptual Mathematics" by Lawvere and Schanuel [172], which is written in a very didactic style and addressed to a general audience, "Category theory" by Awodey [169] and "Arrows, Structures and Functors" by Arbib and Manes [173], which is a strongly recommendable book for beginners in CT with mathematical background. Other gentle introductory textbook on categories written for a non specialist public of mathematically inclined readers are Simmon's "An Introduction to Category Theory" [174] and Pierce's "Basic Category Theory for Computer Scientists [168]" which is a short and clear book addressed to computer science students. Additionally, computer scientists may find in "Category Theory for Computing Science" by Barr and Wells [166] a more advanced exposition of the topic with connections with functional programming languages, and in [175] basic constructions of category theory are written in a functional programming language called ML. The first volume of the Handbook of Categorical Algebra [176] is a credible alternative to Mac Lane's textbook reference. "Categories and Functors" by B. Pareigis introduces category theory from the point of view of representable functors with a notation bit outdated [177]. Finally, students may find useful that both "Introduction to Category theory in for easy movements" by Schalk [178] and [174] include solutions to the exercises proposed.

Now it is time to give a definition of category.

Definition 3.1. A category K is a collection of objects $Obj(K)$, where for each pair of objects A, B of $Obj(K)$, there is a collection of morphisms, $K(A, B)$, being A the domain and B the codomain that satisfies (i) and (ii):

(i) Associativity of composition: for any pair of morphisms $f : A \to B, g : B \to C$, there is a composite morphism, $g \circ f : A \to C$, such that

$$(h \circ g) \circ f = h \circ (g \circ f) \tag{3.1}$$

for all $f : A \to B, g : B \to C, h : C \to D$

(ii) Identity: for each object A, there is a given arrow $1_A : A \to A$ called identity arrow. Thus, for any arrow $f : A \to B$

$$1_B \circ f = f \text{ and } f \circ 1_A = f \tag{3.2}$$

A category is *anything* that satisfies the definition above, the objects can be sets, groups, monoids, vector spaces ...or neurons in the hippocampus. Categories are nothing more than an algebra of arrows equipped with the composition operator.

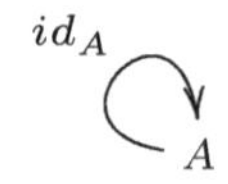

Fig. 3.3 Category **1**

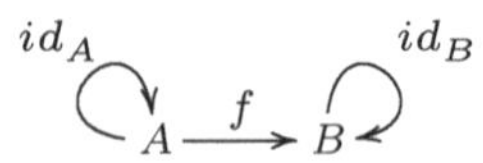

Fig. 3.4 Category **2**

Thus, they provide a domain of mathematical discourse characterized in a very general way [173]. In this vein, classical logic or set theory are particulars in the more general and universal theory of categories.

3.2.1 Examples of Categories

Let us see more examples of categories. Trivially, the category **0** is the empty category with no objects and no morphisms.

The category **1** has one object and one morphism which is necessarily the object's identity arrow (Fig. 3.3).

Category **0** and category **1** are called *discrete categories* because the arrows are the identity arrows, one for each object in the category.

The category **2** has two objects and three morphisms, id_A, id_B and f as the Fig. 3.4 shows.

We first of all must note that the categories shown above – **0**, **1** and **2** – can be expressed as graphs where the objects are nodes and the morphisms are the edges. It is easy to see that every category is a graph (the contrary is not true). Categories **0**, **1** and **2** are small categories because their nodes and arrows constitute sets. A category that is not small is large. For example, the category of sets – whose objects are sets and whose arrows are functions – or the category of graphs – whose objects are graphs and whose arrows are graph homomorphisms – are example of large categories. It ought to be remarked that category theory can handle "the collections of all sets", "the collections of all graphs", "the collections of all groups" etc. and therefore avoids inconsistencies found in set theory such as the Russell's paradox [179] – the set of all sets is not a set.

Now, we provide some examples of categories. First, categories of mathematical structures e.g., the category of sets **Set**, the category of graphs **Graph**, the category of groups **Grp**, the category of monoids **Mon** are defined, and then examples of mathematical structures as categories e.g., pre order sets as a category and natural numbers as a category are given. This distinction between categories of sets with structure e.g. category of graphs and categories with only one set of objects e.g., graph as a category, is of course not the only valid taxonomy of categories but it organizes categories in the way we need in this book. A category **C** is discrete if the only arrows are the identity arrows, that is, $K(A, B) = \emptyset$ for any two objects

$A \neq B$ and $K(A, A) = 1_A$ for each object A. For example, every set $\mathcal{S}$ can be seen as a discrete category **C** whose objects are the elements of $\mathcal{S}$ and the only morphisms are identities. If the objects and the arrows of a category **C** form a set then **C** is a small category or diagram, otherwise it is large. Thus, the category of graphs is an example of a large category and the category of a given graph $\mathcal{G}$ is a small category.

3.2.1.1 Categories of Mathematical Structures

Some classical examples of categories of sets with structure are: the category of sets **Set**, whose objects are sets and whose arrows are functions between sets, the category of pre-order sets **Poset**, whose objects are posets and the arrows monotone functions, the category of monoids **Mon**, whose objects are monoids and the arrows functions that preserve the monoid structure, or the category of vector spaces **Vec**, whose objects are vector spaces and the arrows linear maps. Here we define in detail the category of graphs **Graph** because we will make use of it later in the book. The interested reader may find in the above mention literature the definition of other categories of mathematical structures. In order to define the category of graphs and graph homomorphisms **Graph**, one needs first to define the objects – graphs – and the arrows – graph homomorphisms – of this category.

Definition 3.2. A graph $\mathcal{G}$ consists of

1. A class of objects or nodes $|\mathcal{G}|$ of the graph $\mathcal{G}$
2. A class of arrows $\mathcal{G}(A, B)$ for each pair $(A, B) \in |\mathcal{G}| \times |\mathcal{G}|$

The mapping associated with any two graphs $\mathcal{G}_1$, $\mathcal{G}_2$ such that graph structure is preserved is a graph homomorphism. Thus, graph homomorphisms preserve the abstract shape of a graph. For example graph homomorphisms preserve adjacency.

Definition 3.3. A homomorphism between two graphs $h : \mathcal{G}_1 \rightarrow \mathcal{G}_2$ consists of

1. A mapping $h : |\mathcal{G}_1| \rightarrow |\mathcal{G}_2|$
2. A mapping $\mathcal{G}_1(A, B) \rightarrow \mathcal{G}_2(h(A), h(B))$ for each pair $(A, B) \in \mathcal{G}_1 \times \mathcal{G}_1$

Note also that homomorphisms compose i.e., the composition $h \circ k$ of homomorphisms $h : \mathcal{G}_1 \rightarrow \mathcal{G}_2$ and $k : \mathcal{G}_2 \rightarrow \mathcal{G}_3$ is the new homomorphism $h \circ k : \mathcal{G}_1 \rightarrow \mathcal{G}_3$. Thus, the homomorphic relation on the set of graphs is transitive thanks to the associative property of composition.

Definition 3.4. A walk in a graph $\mathcal{G}$ is a sequence of nodes $n_1, n_2 \dots n_k$ such that n_{i-1} and n_i are adjacent for each $i = 2, \dots k$.

Definition 3.5. A path in $\mathcal{G}$ is a walk in which all nodes are distinct and the length of the path is given by the number of arrows connecting the nodes. Thus, a graph $\mathcal{P}_k$ is a walk that has $k + 1$ nodes $(0, 1, \dots k)$ and k arrows $(01, 12, \dots (k-1)k)$.

Proposition 3.1. *A mapping* $f : |\mathcal{P}_k| \rightarrow |\mathcal{G}|$ *is a homomorphism of* $\mathcal{P}_k$ *to* $\mathcal{G}$ *if the sequence* $0 \xrightarrow{f_0} 1 \xrightarrow{f_1} \dots \xrightarrow{f_{k-1}} k-1 \xrightarrow{f_k}$ *is a walk in G.*

Proposition 3.2. *Graph homomorphisms of $\mathcal{G}$ to $\mathcal{H}$ map paths in $\mathcal{G}$ to walks in $\mathcal{H}$ and therefore do not increase distances. Thus, if $f : \mathcal{G} \to \mathcal{H}$ is a homomorphism then the distance for any two nodes l, m of $\mathcal{G}$ is such that $d_H(f(l), f(m)) \leq d_G(l, m)$.*

Now that we have defined graphs and the precise way in which graph homomorphisms preserve adjacency i.e., graph homomorphisms map paths to walks, we can define the category of graphs **Graph**. It is easy to see that every category is a graph – a category is a graph with a rule for composing arrows – and every functor between categories is a graph homomorphism.

Definition 3.6. A category is a graph $\mathcal{G}$ with two functions $u : C_0 \to C_1$ and $v : C_2 \to C_1$, where C_0 is the set of objects (paths with length 0), C_1 is the set of arrows and C_2 is the set of paths of length 2. The function u is the identity $u(A) = id_A$ for every object A in $\mathcal{G}$ and the function $v(g, f) = g \circ f$ is the composition if arrows f and g are compasable, that is, the codomain of f and the domain of g coincide. The following four properties must hold

1. The source of $g \circ f$ is the source of f and the target $g \circ f$ of is the target of f
2. Composition is associative $(h \circ g) \circ f = h \circ (g \circ f)$
3. The source and target of id_A are both A for every object
4. For every arrow $f : A \to B$, $f \circ id_A = id_B \circ f = f$

3.2.1.2 Mathematical Structures as Categories

The categories seen in the previous section are large categories. To find examples of small categories we just need to think in common mathematical structures such as preordered sets, groups, semigroups, monoids, rings etc. Preorder, poset and monoid can be regarded as categories and they are defined next.

Definition 3.7. A preorder is a set $\mathcal{A}$ equipped with a binary relation $a \leq b$ such that reflexivity and transitivity hold

1. Reflexivity $a \leq a$
2. Transitivity if $a \leq b$ and $b \leq c$ then $a \leq c$

Any preorder $\mathcal{A}$ can be seen as a category **A** by taking the objects to be the elements of $\mathcal{A}$ and taking a unique arrow, $a \to b$ if and only if $a \leq b$. The reflexive and transitive conditions on $\leq$ ensure that associativity and identity rules hold and therefore the preorder category exists.

Definition 3.8. A poset is a preorder plus the condition of antisymmetry i.e., if $a \leq b$ and $b \leq a$ then $a = b$

A category can be seen as a kind of generalized poset. Thus, one can also think of a functor as a generalized monotone map. An example of poset category is the powerset $P(\mathcal{X})$ for any set $\mathcal{X}$ under the usual inclusion relation $U \subseteq V$ between the subsets U, V of $\mathcal{X}$.

Definition 3.9. A monoid is a set $(\mathcal{M}, *, e)$ where $* : \mathcal{M} \times \mathcal{M} \to \mathcal{M}$ and the unit $e \in \mathcal{M}$ hold the following two conditions for all m_1, m_2, m_3

1. $m_1 * (m_2 * m_3) = (m_1 * m_2) * m_3$
2. $m_1 * e = e * m_1 = m_1$

Thus, a monoid is a category with just one object. The arrows of the category are the elements of the monoid, where the identity arrow is the unit element e, and composition of arrows is the binary operation $*$ of the monoid.

3.2.2 *Definition of Some Key Concepts in the Theory of Categories*

In this section I will formally defined some categorical concepts that will be used in this book. Let us start by the concept of diagram. Everyone, by glancing a category theory book, can realise its diagrammatic nature. Diagrams are the way in which this theory proves theorems and states properties of categorical objects. In essence, category theorists do this by exploiting the commutativity of the diagrams. Commutative diagrams work as the equations in category theory, and provide a geometric and therefore more "visual" description than the symbolic insight in the algebraic equations. Let us start by getting acquainted with the "language of category theory", that is to say, diagrams. Diagrams are defined upon the concept of directed graph.

Definition 3.10. A directed graph is a class of nodes, together with a class of oriented edges for each pair of nodes. Thus, a directed graph is a category without composition or identities.

Remark 3.1. A directed graph can be defined without using category theory as a 4-tuple $G = (N; E; \alpha_s; \alpha_t)$ where E is a set of arrows, N is a set of nodes and α_s is a function defined in $E \to N$ that gives the source of each arrow, and the function α_t that gives the target of each arrow. Thus $\alpha_s \circ f = domf$ and $\alpha_t \circ f = codf$.

Definition 3.11. A directed graph in a category **C**, is a diagram whose nodes X_i, X_j ... are labelled by the objects D_{X_i} of **C** and whose edges $X_i \to X_j$ ... are labelled by morphisms in $C(D_{X_i}, D_{X_j})$.

The attentive reader may have realized that according to the above definition, diagrams are graph homomorphisms defined by the mapping D. It is easier to see this with an example.

Example 3.1. Let $\mathcal{G}$ be a graph with objects A, B and C and arrows $f : A \to B$, $g : B \to C$ and $h : A \to C$, as depicted in the Fig. 3.5. The diagraph $\mathcal{D}$ is defined by $D(A) = X$, $D(B) = Y$, $D(C) = Z$, $D(f) = u$, $D(g) = v$ and $D(h) = w$, in Fig. 3.6.

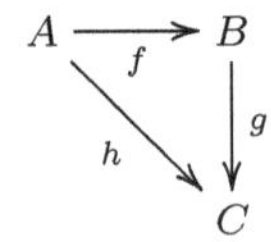

Fig. 3.5 Graph $\mathcal{G}$

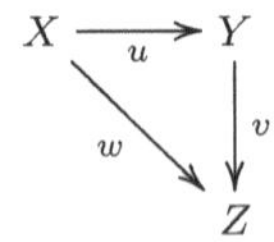

Fig. 3.6 Diagraph $\mathcal{D}$

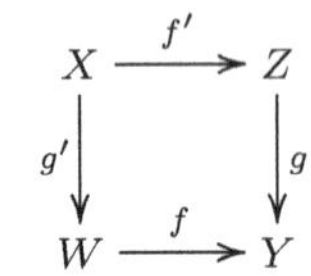

Fig. 3.7 Commutative diagram

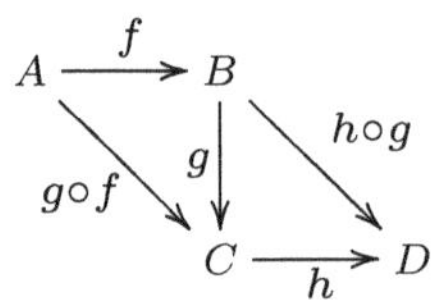

Fig. 3.8 Diagram with associativity of composition

Definition 3.12. A diagram in a category **C** is commutative if for every pair of nodes X, Y in the diagram, all the paths in the diagram yield the same arrow in the category **C**.

Thus, the path in the diagram $X \to Z \to Y$ which is the arrow $g \circ f'$ in the above Fig. 3.7 is equal to the path $X \to W \to Y$, which is the arrow $f \circ g'$ in **C**. Note that the two paths between X and Z compose the same arrow $g \circ f' = f \circ g'$, because for every argument they have the same value. For example, the associativity of composition that all categories must have based on the definition of category (Definition 3.1), means that in the diagram depicted in Fig. 3.8, the two paths or composition of arrows, $(h \circ g) \circ f$ and $h \circ (g \circ f)$ are identical, $(h \circ g) \circ f = h \circ (g \circ f)$.

The notion of dual is very important on category theory, because all the definitions in category theory can be dualized, that is to say, the definitions come in pairs, which obviously simplifies the work of defining and understanding concepts in a theory. Some examples of dual objects are initial object/terminal object, product/coproduct, limit/colimit, cone/cocone, monomorphism/epimorphism etc. The dual of a category **C**, $\mathbf{C}^{op}$, can be obtained by reversing the arrows. To illustrate this point the definition of dual category and cone and cocone are provided next.

Definition 3.13. The dual **C** of a category $\mathbf{C}^{op}$, is such that the objects are the same, $|K| = |K^{op}|$, and there is a one-to-one correspondence between the arrows i.e., for every arrow $f : A \to B$ in **C** there is a dual arrow $\bar{f} : B \to A$ in $\mathbf{C}^{op}$ with composition and identity properties defined in the normal way.

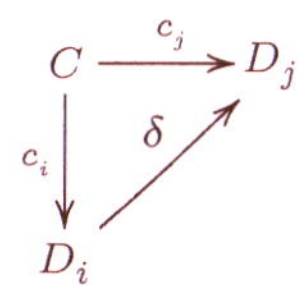

Fig. 3.9 C is a cone for the diagram D

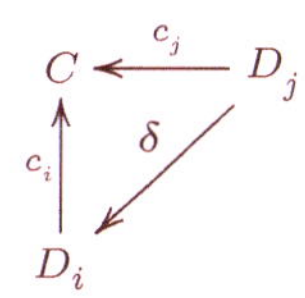

Fig. 3.10 C is a cocone for the diagram D

Definition 3.14. A cone to a diagram D is an object C, and a family of arrows $c_i : C \rightarrow D_i$, such that for every arrow $\delta : D_i :\rightarrow D_j$ in D, the diagram in Fig. 3.9 commutes. C is called apical object and the morphisms $c_i : C \rightarrow D_i$ leg morphisms (Fig. 3.9).

The cones for D form a category of cones **Con(D)**.

Definition 3.15. The dual of a cone is a cocone. Thus a cocone of a diagram D is an object C and a family of arrows $c_i : D_i \rightarrow C$, such that for every arrow $\delta D_j \rightarrow D_i$ in D, then $c_i \circ \delta = c_i$ which is equivalent to the predicate that the diagram in Fig. 3.10 commutes.

3.2.2.1 Universal Constructions in Categories

Category theory starts with the realisation that most constructions in maths may be defined by universal properties. The basic notions of category theory are six: object, morphism, domain, codomain, associativity and identity. However, strictly speaking, there is only one: morphism, because the other five: object, domain, codomain, identity and associativity, rely on the concept of morphism. Furthermore, following Kan and Grothendieck investigations, the concepts of object, morphism, domain and codomain can be described in terms of adjoint functors [180]. Here we will focus on first order concepts of category theory, thats is to say, concepts that strictly belong to categories and functors. Higher order relations like natural transformations and adjoint functors will be defined later, when needed in Sect. 3.4. Before starting with the survey of universal constructs, we need to be precised about what is intended here by universal. A universal construction describes a class of objects and accompanying arrows that share a common property. The universal properties allow us to do not have to deal with all the particular details of a construction, but to prove if the construct verifies a universal property. If that is the case, the details render unnecessary because all what there is to know about the construct is already contained in the universal property. This is in actual fact at the basis of all scientific modelling.

The rationale behind universal properties is to obtain definitions of objects as general and descriptive as possible. Category theory can encompass these two apparently divorced terms – generality and descriptiveness – because it allows to capture the structure of the object in the most general way. For example, it has been suggested that universals of the mind may be expressed using universal properties in the theory of categories [181]. The universal properties monomorphism, epimorphism, isomorphism, initial and terminal object, product and coproduct, limit and colimit are defined next. The section concludes with the definition of functor.

Definition 3.16. A morphism $f : A \to B$ in a category K, is a monomorphism if for every pair of arrows $g, h : X \to A$, $f \circ g = f \circ h$ then $g = h$. Diagrammatically, f is a monomorphism if whenever the following diagram commutes implies $g = h$.

$$X \underset{h}{\overset{g}{\rightrightarrows}} A \xrightarrow{f} B$$

Remark 3.2. Note that the predicate "for every pair of arrows $g, h : X \to A$, $f \circ g = f \circ h$ then $g = h$" states an universal property.

Definition 3.17. A morphism $f : A \to B$ in a category K, is an epimorphism if for every pair of arrows $g, h : B \to X$, $g \circ f = h \circ f$ implies $g = h$. Diagrammatically, f is an epimorphism if whenever the following diagram commutes, implies $g = h$.

$$A \xrightarrow{f} B \underset{h}{\overset{g}{\rightrightarrows}} X$$

Definition 3.18. A morphism $f : A \to B$ in a category K, is an isomorphism if there is a morphism $g : A \to B$, such that $g \circ f = 1_A$ and $f \circ g = 1_B$. The objects A and B are said to be isomorphic because it exists and isomorphism between them, and since inverses are unique, g is the inverse of f, $g = f^{-1}$. Trivially, an isomorphism in K is also an isomorphism in K^{op}.

Definition 3.19. An initial object of a category K is an object 0 in $|K|$ such that for any object X in $|K|$ there is one and only one morphism $0 \xrightarrow{f} X$.

Thus, the idea behind initial object is the smallest thing with all the structure in question. For example, in the category of sets **Set**, the empty set $\emptyset$ is a terminal object. In the category determined by one poset,[1] an initial object is the absolute minimum for the poset (in the case it exists).

The concept of terminal object is dual of initial object.

Definition 3.20. A terminal object of a category K is an object 1 in $|K|$ such that, for any object X in $|K|$ there is one and only one morphism $X \xrightarrow{f} 1$.

[1]Note that the category determined by one poset is not the same than the category of posets **Pos**

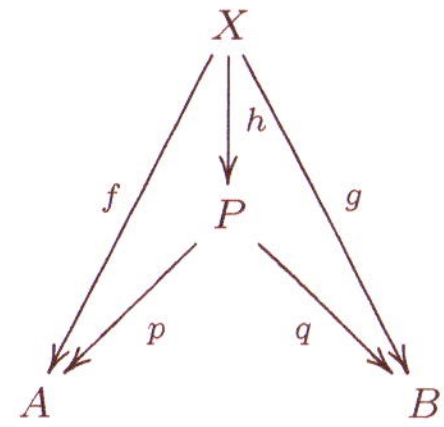

Fig. 3.11 P is the product of A and B

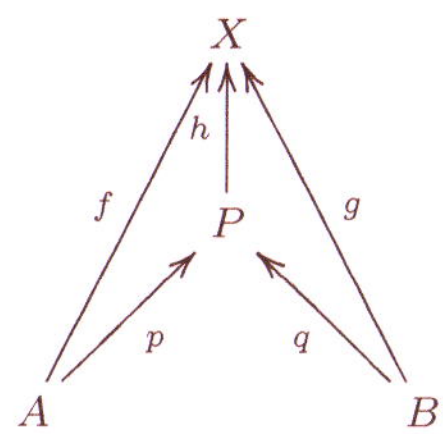

Fig. 3.12 P is the coproduct of A and B

Proposition 3.3. *Initial (terminal) objects are unique up to isomorphism.*

Proof. Given that C and C' are both initial objects, the following diagram shows that C and C' are uniquely isomorphic.

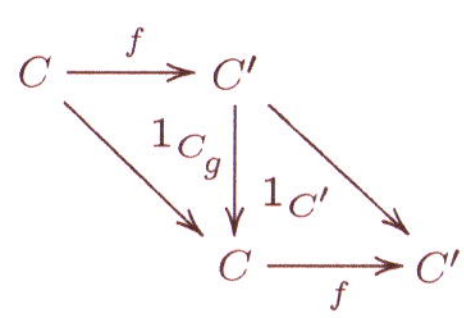

The demonstration for terminal object is identical using the dual category. In the category determined by a poset the terminal object is the absolute maximum for the poset (in the case it exists).

Definition 3.21. A product of two objects A, B in $Obj(K)$ is an object P equipped with two morphisms, $P \xrightarrow{p} A$ and $P \xrightarrow{q} B$, such that given any pair of morphisms, $X \xrightarrow{f} A$ and $X \xrightarrow{g} B$, there is a unique morphism h making the following diagram commutes, Fig. 3.11.

A coproduct is the dual of product, so it can be obtained reversing the arrows in the above diagram. Formally,

Definition 3.22. A coproduct of two objects A, B in $Obj(K)$ is an object C with two morphisms, $A \xrightarrow{p} P$ and $B \xrightarrow{q} P$, such that for every object $Obj(K)$ and all pairs of morphisms, $A \xrightarrow{f} X$ and $B \xrightarrow{g} X$, there is an unique morphism h making the following diagram commutes (Fig. 3.12).

The definition of coproduct can be generalised to n representations as follows.

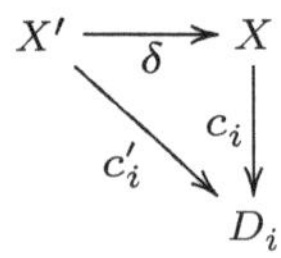

Fig. 3.13 The cone $(c_i : X \to D_i)$ is a limit of the diagram D

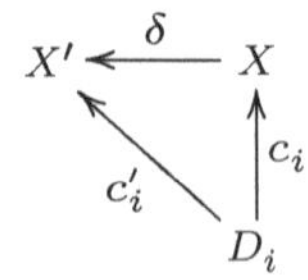

Fig. 3.14 The cone $(c_i : D_i \to X)$ is a limit of the diagram D

Definition 3.23. A coproduct of a family of n objects Ai consists of an object $\coprod_i A_i$ together with a family of arrows $(A_i \xrightarrow{\pi_i} \coprod_i A_i)$ such that for any object C and family of arrows $(A_i \xrightarrow{\alpha_i} C)$ there is an unique arrow $(\coprod_i A_i \xrightarrow{h} C)$ that makes the following diagram to commute.

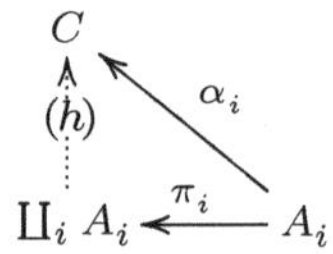

As it will be shown in Sect. 3.4.4.3, Philips et al. [182] have pointed out that fundamental cognitive principles such as transitive inference or class inclusion, involve the capacity to compute coproducts. Products and coproducts can be defined in a more general way. Indeed, a product is a form of limit and coproduct is a form of colimit.

Definition 3.24. A limit for a diagram D is a terminal object in $Con(D)$. Thus the cone $(c_i : X \to D_i)$ is such that for every cone in D, $(c_i^{'} : X' \to D_i)$, there is an unique morphism $\delta : X \to X'$ such that the triangle in Fig. 3.13 commutes.

Definition 3.25. A colimit for a diagram D is a cocone with the universal co-property that for any other cocone there is a unique arrow such that the diagram in Fig. 3.14 commutes. Initial objects are colimits of empty diagrams (Fig. 3.14).

The importance of limits and colimits relies on the fact that they are general mechanisms for combining structures. For example, in [183], Ronald Brown and Timothy Porter suggest that colimits may give useful analogies to understand the way in which complex systems operate. The concept of colimit may help to explain how individual parts provide a working whole, that is to say, it captures the structure of the system, making irrelevant the choice at intermediate stages in the dynamics of the system. Ehresmann and Vanbremersch have been pioneers in finding the relevance of colimits to biological development [184]. The dynamics of the category of neurons produces evolving structures which can be captured with colimits. This notion of colimit allows to "put together" different neural process into a coherent whole. In Chap. 8, colimits are presented as the main modeling tool for spatial representation in the brain hippocampus. There, it is described the effect in injecting the concepts of co-product and colimit from category theory into the problem of place cell formation in the hippocampus.

As noted above, links between objects in a category are called morphisms or arrows. But there are also exist links between categories, these are called functors. A functor is a structure preserving map between categories, that is, as an arrow in the category of categories, which maps the domain category onto the codomain category. Let **A** and **B** be categories, a functor $F : \mathbf{A} \rightarrow \mathbf{B}$, maps every object A of category **A** to an object $F(A)$ in **B** and each morphism in **A** to those of **B**. Formally,

Definition 3.26. A functor $F : \mathbf{A} \rightarrow \mathbf{B}$, from category **A** to category **B** verifies:

1. A mapping of objects to objects, $F : |\mathbf{A}| \rightarrow |\mathbf{B}|$, such that for all $A \in |\mathbf{A}|$, $F(A) \in |\mathbf{B}|$
2. A mapping of arrows to arrows, for each arrow $f : a \rightarrow a', a, a' \in \mathbf{A}$ exists a morphism $F(f) : F(a) \rightarrow F(a') \in \mathbf{B}$ such that for all $a \in |\mathbf{A}|$ and all composable arrows $f, g \in \mathbf{A}$:
   (a) $F(1_a) = 1_{F(a)}$
   (b) $F(g \circ f) = F(g) \circ F(f)$

The composition of two functors $F : \mathbf{A} \rightarrow \mathbf{B}$, $\mathbf{B} \rightarrow \mathbf{C}$ gives a new functor $\mathbf{A} \rightarrow \mathbf{C}$. The demonstration is straightforward: the identity functor in the category **A** (define every mapping as the identity) is an identity of that composition law which is obviously associative.

Example 3.2. Let **C** be any category of structured sets, for example the category of monoids. Thus each object in **C** is a monoid, then by the definition of monoid (Definition 3.9) each object is a set, and each arrow or monoid homomorphism is a function between two underlying sets. Thus, $F : \mathbf{Mon} \rightarrow \mathbf{Set}$ is a functor that takes every monoid to its underlying set and every monoid homomorphism to its underlying function. This is an example of forgetful functor because some structure is "forgotten" as we pass from the domain category (**Mon**) to the codomain category (**Set**)

Example 3.3. If one "forgets" that in categories arrows can be composed and which arrows are the identities, then the structure that is remember is that of a graph. This gives the forgetful functor $U : \mathbf{Cat} \rightarrow \mathbf{Graph}$ since every functor is a graph homomorphism but not every graph homomorphism is a functor.

Other examples of forgetful functors are $F : \mathbf{Rng} \rightarrow \mathbf{AGrp}$ the functor from the category of rings to the category of abelian groups in which the x-structure of rings is forgotten, and $F : \mathbf{Grp} \rightarrow \mathbf{Mon}$ between groups and monoids where the inverse property of groups is forgotten. Thus, we have a forgetful functor $F : \mathbf{A} \rightarrow \mathbf{B}$ when it is possible to send every object and every arrow in **A** to its respectively underlying object and arrow in **B**. Identity and composition are obviously preserved.

Like the functions that map sets to sets in the category **Set** there are arrows between categories called functors, and there are further mappings between functors called natural transformations. Thus, categories compare two objects, functors compare two categories and natural transformations compare two functors.

$$Source \underset{G}{\overset{F}{\rightrightarrows}} Target$$

Fig. 3.15 Functors F and G

$$FA \xrightarrow[\tau A]{} GA$$

Fig. 3.16 F A and G A are objects of Target. A natural transformation is a family of arrows of Target indexed by the objects A of Source, τ_A

$$\begin{array}{ccc} FA & \xrightarrow[\tau_A]{} & GA \\ {\scriptstyle F(f)}\downarrow & & \downarrow{\scriptstyle G(f)} \\ FB & \xrightarrow[\tau_B]{} & GB \end{array}$$

Fig. 3.17 For each arrow $f : A \rightarrow B$ of Source the square in Target commutes

Definition 3.27. Given a pair of functors F and G, as in Fig. 3.15, a natural transformation is a set of arrows in the category **Target** indexed by the objects A of the category **Source**, as in Fig. 3.16, such that for each arrow $A \rightarrow B$ in **Source** the diagram depicted in Fig. 3.17 commutes.

The two functors F and G produce two pictures of **Source** inside **Target**. For an object a in **Source**, this gives two objects, $F(a)$ and $G(a)$, in **Target**. A natural equivalence is a natural transformation for which the arrows are invertible, that is, τ is an isomorphism.

$$F \xrightarrow[\tau]{} G$$

This is similar to the relation between category and groupoid, a category whose arrows invert. In the progression from set to category to functor, equality between members of the set has passed to equivalence between categories and thence to isomorphism between functors. This is isomorphism in the sense that the invertibility of operations induces a correspondence analogous to "equality". The hierarchy has a natural continuation: functors (2-categories) generate natural transformations (3-categories). This iterative process may continue until a (n + 1)-category i.e., a collection of all n-categories with objects, morphisms, 2-morphisms, ... up to n-morphisms.

3.3 The Cat-Level Avenue

The brain is an ultra complex system with the additional difficulty that its behavior is extremely complicated to analyse in vivo, requiring elaborated experimental procedures and expensive technical equipment. Although the trees are well known – detailed models at microscopic level – there is still missing an unified theory of the wood – macroscopic models. Unfortunately, it seems that in brain sciences, descriptive understanding is divorced from the understanding of underlying mechanisms. Put it in a different way, explain the higher level properties like for example: memory or language, in terms of lower level properties, like action potential of firing rate of single neurons, requires far more than statistical inferences, analogies, metaphors or intuitions. But there are still good news, the brain dynamics is, at least in principle, amenable to mathematical representation.

In Chap. 2, it has been described a number of different mathematical tools applied to neuroscience. Depending on the level of organisation in the brain, we find different approaches and techniques better suited than others. We thus, have a whole host of mathematical structures and methodologies to integrate within a general framework. In order to acquire a global understanding of brain dynamics, we need to develop new methodologies able to fuse, or at least vertebrate, all these different approaches. It might be remarked here that I am not claiming for an unified theory of cognition of any sort, but to bring in a new methodology, able to modify in a considerable manner the existent frameworks so allow for new results to be obtained. This approach is based on the mathematical theory of categories and tries to relate different levels, as well as making clearer the mathematical structures needed at each level, and their efficiency in obtaining relevant explanations of what is observed in the brain.

This is in a nutshell how category theory cope with the hierarchy of organisational levels:

1. Detect the localisation of the meaningful patterns of components at $level_i$
2. Build the colimit that embodies an object at $level_{i+1}$. The colimit is obtained as a result of the binding process of components at level $level_i$ and is a category of objects of $level_i$
3. Go to step 1 in order to continue the construction of categories of higher level i + 2 ...

The mathematical theory of categories may help to set the basis of a novel framework to cope with the phenomenon of emergence in neuroscience. In particular, it gives a common framework for studying the different levels of description. In neuroscience, these levels are typically assumed to be two: microscopic and macroscopic, or three: microscopic, macroscopic and the mesoscopic level, which lies between those. There are surely many more levels, each feeding into and being controlled by networks at other levels. We need mathematical objects able to deal with these networks. However, from the microscopic level $level_1$ where the category of neurons, **Neur** – whose objects are neurons and the morphisms are the synaptic paths between them, with the convolution of paths as composition – is placed all the way up to the $level_N$ of mental objects captured by the category **ImO** – whose objects are mental images and the morphisms are relationships between them e.g., is part of – there are $N-1$ levels that need to be addressed. All those $N-1$ levels are in truth multiple scales of organization that go from local network of neurons to the scalp topographic scale and the statistical correlation

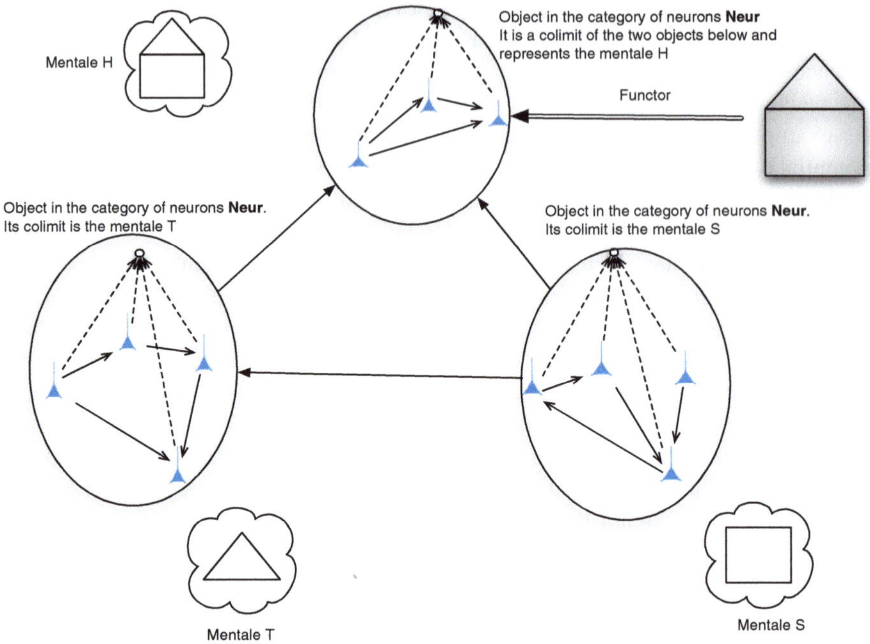

Fig. 3.18 The figure depicts the categorical construction of a physical object, the *gray* house on the *right*. The **Mentalae** S and T on the *bottom* of the figure are concepts reified in a neural organisation, the category **Neur**. The colimit of the objects represent the *Mentale* H (the house). Finally there is a functor between the physical object and the neural category **Neur**

of time series of for example, electroencephalography and functional magnetic resonance imaging. In this vein, the category of neurons **Neur** and the category of mental objects **ImO** are models of the brain at very different hierarchical levels, namely the microscopic and macroscopic. Thus, the category **Neur** is composed of neurons and models the physical structure of the brain and its elementary neural dynamics. Figure 3.18 exemplifies this.

Category theory may reframe the fundamental question of how cognitive functions, including higher mental processes, arise from the structural connectivity of the brain. Categories of increasingly complex objects, are built and formally specified, using terms like for example product, colimit or functor. In [185], E.C. Zeeman pointed out that a mathematical explanation of how the brain works has to rely on the concept of isomorphism.

> First develop a piece of mathematics X that describes the permanent structure (memory) and the working (thinking) of the mind, and another piece of mathematics Y that describes the permanent structure (anatomy) and working (electromechanical) of the brain; then, from hypothesis based on experimental evidence, prove an isomorphism $X \cong Y$

Of course, we must be cautious with the above referred *experimental evidence* in brain studies mentioned in an article written in 1962. However, the approach and the emphasis given to the concepts of structure and morphism are totally relevant. There are two main issues to be considered from the above quote. First, if category theory is the mathematical theory that studies the structure, we are going to necessarily need such theory or theories alike when we try to model the structure and function of the brain. Importantly, the focus is not in what the objects may be or do, but on the characterization of the arrow-patterns between the objects. The arrows reveal the structure and any universal that pervade the structure. Second, the relation of representation between two systems involves the establishment of a mapping, ideally an isomorphism, or in the actual case, a homomorphism, as it will be shown in Chap. 6. In category theory it is possible to define morphisms between objects within a category, but also, at the immediately superior level, we can find morphisms between categories – these comprise functors – which are structure-preserving maps between categories as previously defined in Definition 3.26.

3.4 Applications of Category Theory in Cognitive and Brain Science

This section's rationale is to review some of the most relevant applications of the mathematical theory of category in biology, cognitive psychology, neuroscience, semiotics and knowledge representation [186].

There is a long and fruitful tradition of using category theory in informatics and specifically in software engineering. For example, in programming language semantics, modular specification or formal software validation. It would take too long to include that relation here, for those interested in category theory in computer programming can see [187] and references therein. It is worth mentioning that programming languages such as Huskell [188], and up-and-coming languages like SCALA [189] or Clojure [190] contain libraries to implement objects of the theory of categories It might be remarked as well that at the moment of writing this, 2013, category theory is not a standard mathematical tool in any of the empirical sciences. The only exception to this can be found in quantum mechanics, where category theory is a well known theory for those working in quantum gravity and related fields [191]. However, quantum mechanics cannot be considered an empirical science *tout court*, empirical testing in quantum gravity is still an ongoing process. The construction of large energy particle accelerators will help in this direction.

3.4.1 The Origins: Rosen's (M,R-Systems)

The application of category theory to biological systems began as early as the 1950s, with the attempts of Robert Rosen within the frame of relational biology. Rosen's work is often cryptic, which unfortunately has helped to portray the idea that it is indeed a very interesting formalism but too abstract and hard to understand, let alone to apply. Rosen's theory aims to transcend the reactive paradigm of simple physical systems to a new post-Newtonian paradigm, able to deal with complex natural systems, or organisms using Rosen's terminology [192].

The metabolic, repair systems, (M,R)-systems, is a way of characterising systems quite independently of their physic-chemical constitution. In Rosen's words

> What remains then is an abstract pattern of functional organization, which has properties of its own, independent of any particular way it might be materially realized. Indeed, it is what remains invariant in the class of all such material realizations, and hence characterizes that class. [193]

The theory of categories provides the mindset to model the (M,R)-systems which are a kind of systems that are independent of how they are physically realized. The (M,R)-systems characterise the minimal organisation a material system would have to manifest in order to realise a cell. Rosen acknowledged, of course, that there exist also in biology, systems that depend upon the material details of a particular realization. In order to represent (M,R)-systems it is needed a general class of diagrams that is provided by the theory of categories. Rosen was the first in seen the power of category theory to express, in a purely mathematical realm, patterns of relations between models and between models themselves. This fact was exploited by Rosen who was looking for a theoretical framework for modeling in general, and in particular for living systems.

In essence Rosen's theory is as follows. A completely autonomous living organism needs to encode all the information about the state of all of its catalysts (the M in (M,R)-system), and to make, when necessary, the necessary replacements itself (the R in (M,R)-system). The only way this information can be encoded, is in the connectivity of the network itself. Thus, the network must be constructed in such a way that its entire connectivity landscape must be uniquely encoded i.e., the natural system – the living organism – and the formal system – the model of the organism – are isomorphic [194]. The formal system is relational, consisting of a category of interrelated components. A component is defined by the mapping $f : A \to B$. For example, f can represent an enzyme that converts a substance A into a product B. Both metabolism (M) and repair (R) processes are modeled as morphisms, $M : A \to B$ and $R : A \to H(A, B)$ respectively, where H represents the class of all the metabolisms from A to B or $M : A \to B$. Rosen, with the (M,R)-systems mechanism tries to provide a solution to the infinite regression in the iterative mappings M,R and H. They key issue is how the network produces the selector map g in order to avoid that vicious circle. Graphically,

$$A \xrightarrow{f} B \xrightarrow{g} H(A, B)$$

where the operator $g : B \to H(A, B)$ is produced by the network when the system is capable of autonomous invariance.

The mapping

$$\beta : H(A,B) \to H(B, H(A,B))$$

encapsulates the notion of metabolic enclosure and provides the solution to the vicious circle reflected in the recursive map β. For β to exist, given a metabolism f, the corresponding selector g must be such that $g(f) = f$.

$$A \xrightarrow{f} B \xrightarrow{g} H(A,B) \xrightarrow{\beta} H(B, H(A,B))$$

3.4.2 Category Theory in Perception

In [195] Z. Arzi-Gonczarowski deploys a basic category theoretic tool for perception modeling purposes. A perception is defined as a 3-tuple $< E, I, \rho >$ such that E and I are finite disjoint sets and ρ is the arrow $\rho : E \times I \to \{t, f, u\}$. The set E is made of elements of the external world, and I are mental concepts or internal connotations of the cognitive agent. Therefore the predicate $\rho : E \times I \to \{t, f, u\}$ is 3-valued. A mental concept $i \in I$, can be a true, a false or an unknown connotation of an external object $e \in E$.

Let E be an environment and $P_1 =< I_1, \rho_1 >$ and $P_2 =< I_2, \rho_2 >$ two perceptions over E. The mapping $h : P_1 \to P_2$ is a perception morphism (p-morphism) iff h is a mapping between the connotations I_1 and I_2 and definite truth values (t,f) are preserved by the p-morphism. Note that p-morphisms are the categorical morphisms of the category based on the collection of all perceptions with the same environment E.

3.4.3 Memory Evolutive Neuronal Systems

In order to understand the theory developed by the mathematician Andree Ehresmann and the physician Jean-Paul Vanbremeersch, it is necessary to understand first the approach to systems modeling based on category theory in which the theory is rooted. In Chap. 2 it was introduced the concept of state space, which is the space with all possible states of a system. The state space approach is non structured in the sense that a state space is merely a collection of points. For example, the state of a pendulum is given by the quantities representing the position and the velocity like in $\{(0,0), (pi/4, 0.1), (pi/2, 0.3)\}$. This is in contrast with a categorical configuration approach which imposes the structure of categories to the system's state representation. The identity, and very importantly, the composition which, according to Healy and Caudell [17], is the most important distinguishing feature of a category are here taken into account. The rationale behind the categorical approach is to model the system's dynamics using categories and functors. Thus, the successive configurations of a system are captured by categories, and changes of configuration by partial functors between categories. In this view, it is always possible to characterised any given system as a graph of interdependent objects (nodes) with relations between them (edges). These objects and relations together constitute the internal organization, or configuration of the system under study at a particular time.

We are interested in the relations between configurations at different times, that is, relations between two graphs such that the same basic structure is shared. Let G_1 and G_2 be any two graphs, a homomorphism of G_1 to G_2, $f : G_1 \to G_2$ is a mapping such that for every node in G_1 there is a node in G_2, $f : |G_1| \to |G_2|$ and to every arrow $u \circ v$ in G_1, $f(u) \circ f(v)$ in G_2. Thus, graph homomorphisms are assignments of the nodes of one graph to the nodes of another graph, satisfying the adjacency constraints. The homomorphisms among a set of graphs impose the algebraic structure of a category. Every finite category can be represented by a set of graphs, where each graph is an object for the finite category. The transition between different system configurations is based on the fact that homomorphisms compose. It is important to note that composition of homomorphisms is a partial operation i.e., it can only be defined if the codomain of the first arrow is the domain of the second arrow. The composition of automorphisms (an isomorphism $f : G \to G$), and endomorphisms (an homomorphism $f : G \to G$) are always defined. The point we must retained here is that categories are richer structures than graphs, and graph homomorphisms within a category preserves adjacency which is a kind of transitivity of relations.

A case in point of the multiple ways in which this categorical setting offers new perspectives for modeling complex systems is the theory of Memory Evolutive Systems (MES). MES produces a mathematical model for natural open self-organizing systems, such as biological, sociological or neural systems. In particular, the theory has been applied to neural systems i.e., Memory Evolutive Neural Systems (MENS). MENS is a particular case of MES. Thus, MENS is a mathematical model based on the theory of categories for cognitive systems, and MES is a more general model that deals with problems like emergence and self-organisation for autonomous complex hierarchical systems in a formal-categorical way. In this section we focus on MENS.

As it has been already said, category theory is based upon the notion of morphism or arrow. This is in contrast, for example, with set theory which relies in the concept of membership of a quantity x, in a collection or set X, $x \in X$. A morphism, for example, $f : a \to b$ in a category C, expresses one of the many possible ways in which the object a relates to the object b in the context given by the category C. We can easily find the underlying graph G provided by the morphisms in a category C. The objects in C are the nodes in G and the morphisms are the edges in G. However, it might be emphasized that a graph and a category are not the same thing. In fact, a category can be seen as a (multi)-graph but equipped with something more: a composition law. In a category, compositions are calculated when two morphisms have identical domain and codomain. Category theory exploits this symmetry by stating that different paths whose compositions have the same domain and codomain may have the same meaning.

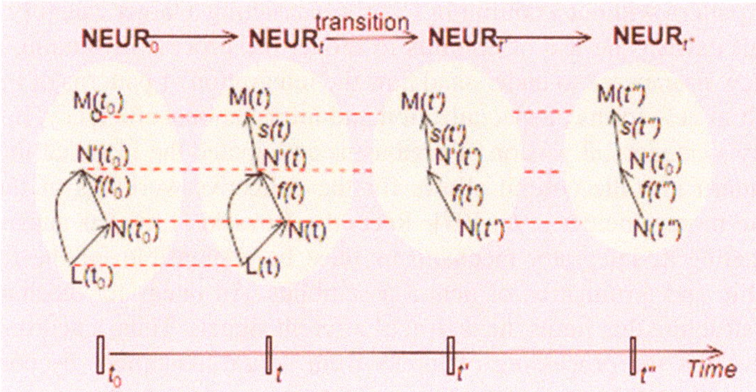

Fig. 3.19 The figure describes the evolution of the neural system during its life. For each t of the timescale we have the category $NEUR_t$ of neurons at time t (Figure courtesy of Andree C. Ehresmann)

MENS starts with a multi-directed graph that models a nervous system, which is the lower model in the hierarchy of categories that is constructed in this theory. At such level, a graph will map the neural assembly as follows: for every real neuron, a node is created in the graph, and for every synapse between two neurons, there is an edge in the graph that connects the two nodes that represent the presynaptic and postsynaptic neurons. A sub-system **Neur** is introduced in the theory to model the neuronal system. Of course, the nervous system is dynamic or evolutive in Ehresmann's words, and neurons and synapses may be created or disappear.

We can now start to model the system using the categorical approach that was referred above. The configuration of the neural system at a given time t is given by the category of neurons $\mathbf{Neur}_t$, and the trajectory of the configuration from the instant t to another instant t', is given by the partial functor $F : Neur_t \to Neur_{t'}$, which maps the configuration of neurons and synapses existing in **Neur**, at two different times points (Fig. 3.19).

Now, we need to find the relevant patterns for **Neur**. A synchronous assembly of neurons P may work as an informational input on another neuron N. In this case, we say that N is a coordination neuron whose activity corresponds to the synchronous assembly of neurons P. It may happen that N cannot be found, but there may exist a colimit **cP** at a superior level than that of **Neur**, that is to say, a cat-neuron **cP** of an immediately superior level may emerge. In that case, the cat-neuron **cP** becomes the colimit of P and also of other patterns P' synchronised with P. Thus, MENS begins with a sub-system **Neur** that is initially modeled as a graph, which will suffer transformations that will be modeled through partial functors in the configuration category space, given by the category of neurons at different given instants $\mathbf{Neur}_t$. The system will evolve, and by the complexification process, higher cat-neurons representing more complex brain processes will rise. The complexification process in a category **C**, as defined in [184], consists in binding a pattern without a colimit in **C**, by constructing a larger category in which the colimit exists. The use of colimits to model the process of binding a pattern offers a new framework to understand how the integration of patterns of interaction may result in new forms of specialization and informational efficiency. In this line, Rashevsky's categorical version of organisms anticipated the idea that the binding of the pattern into the colimit allows for the collective workings of the pattern to become more efficient [196, 197]. Recently, Buzsaki [115] has suggested that neurons acting as integrator mechanisms must be responsible for the formation, membership and termination of neural assemblies. An integrator mechanism is a neuronal structure that limits the arrival of afferent signals. Thus, a neuron or group of neurons bias the processing of inputs from neural assemblies by consistently constraining the afferent information in a certain way. Much as there is not a single neural code, the integrator mechanism is not unique, as it varies depending on the level of analysis. An interesting possibility worth exploring is to model this integrator mechanism as a colimit in the category of neurons.

3.4.3.1 The Binding Problem with a Categorical Outlook

It is important to remark that a cat-neuron in MENS is conceived as a multifold object because it admits several decompositions patterns $P, P' \ldots$ which may be not connected. This point is crucial because it establishes a powerful mathematical tool to explore the *binding problem:* or how different functional brain areas for particular items, are coordinated in such a way to form widely distributed assemblies of neurons that represent complex items. Using Ehresmann's words "the cat-neuron characterizes the invariant that the different assemblies it binds have in common [184]". In this light, the cat-neuron colimit is a formalisation of the multiplicity principle, which is a generalisation of what neurobiologists call the degeneracy problem in neuronal code – given a single group of neurons they may participate in the coding of more than one function. The multiplicity principles will be fully explored in Chap. 4. This is roughly how the complexification of a category of neurons is formed using the recursivity property of category theory.

1. Build the subsystem **Neur** which is represented as a graph having as nodes neurons and as edges synapses. The cat-neuron of level 0 is formed.
2. Synchronous assemblies of neurons form patterns and a cat-neuron of level 1 is created when a colimit to those patterns can be found.
3. By induction a cat-neuron of level k is the colimit of a series of patterns of lower cat-neurons$_{k-1,1}$.

As Fig. 3.20 shows, the binding of a pattern P of neurons in a category of level 1 (**Neur**), becomes the mental image ImO of an object O. Thus, ImO is a *cat-neuron* of level 1. Progressively, the construction of a cat-neuron of higher level, 2 and so on, is established by the mental image of an object C formed by the juxtaposition of several objects O_i that the cognitive system e.g., an animal can already recognize. The fundamental question of *how higher mental processes arise*

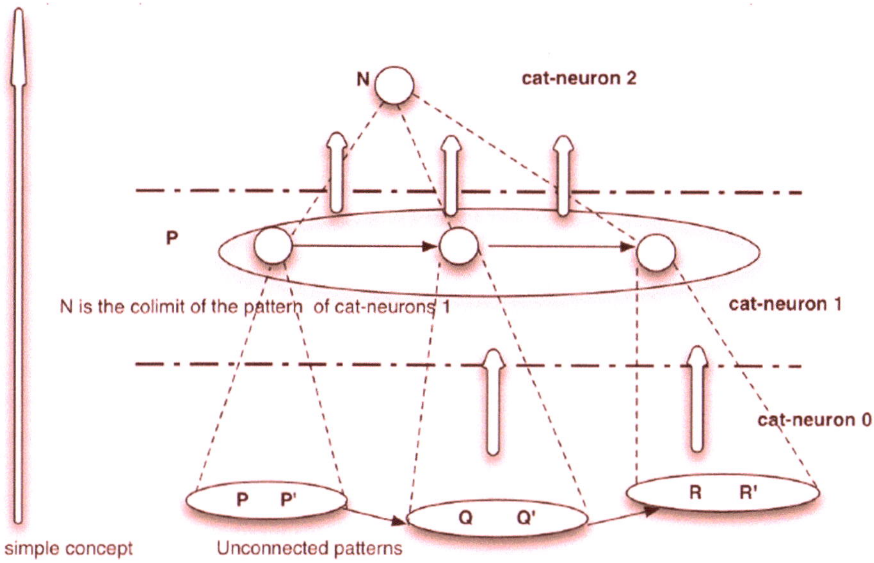

Fig. 3.20 The iterative building of cat-neurons binding patterns of cat-neurons of lower levels, models the computation on mental objects to form more complex ones. It gives explanatory power for the construction of an algebra of mental objects

from the functioning of the brain? is approached by the formation of increasingly complex objects. In this vein, the categories **Neur** and **ImO** of neurons and mental objects are models of the brain at different hierarchical levels.

MENS does not explain how the patterns of neurons are formed or how the category of neurons of higher complexity are created. However, it does provide fresh theoretical insights that deserve to be explored by either empirical tests or computational simulations. Indeed it tackles fundamental conceptual issues like, the phenomenon of emergence or the degeneracy principle in biological systems, and it does so by formulating new mathematical structures, adapted to the structure and functioning of the brain/mind. What is necessary now is for experimentalists to validate the theoretical hypotheses by measuring neural and cognitive activity. Only then, will it be possible to establish the real value of the theoretical claims and the mathematical structures proposed in MENS. This book intends to make progress in that direction, and in Chap. 8 a testable theory that is able to explain how place cells in the hippocampus are created will be described.

3.4.4 *Category Theory in Knowledge Acquisition and Representation*

In [180], Lawvere points out that category theory will be a necessary tool in the construction of an adequately explicit science of knowing. Here we investigate

```
Concept T0
  sorts Points, Lines
  op on: Points*Lines →Boolean
  Axiom Two-points-define-a-line is
    forall(x, y:Points) ((x not = y) implies
      (exists L:Lines) (on (x, L) and on (y, L) and
        ((forall m:lines) (on (x, m) and on (y, m))
          implies (m = L) ))
end
```

Fig. 3.21 The theory T0 defines a simple geometry of points and lines which are primitive types and the relationship between them is given by the axiom "Two-points-define-a-line" [17]

how this idea has progressed since Lawvere's claim. We will see how "the initial plausability" claimed by Lawvere has been transformed into real implementations, for example, in the construction of ontologies using adaptive neural networks (Sect. 3.4.4.1), in categorical models for family resemblance and contextuality as well as in semiotics (Sect. 3.4.4.2), or in the field of cognitive psychology, for example in concept formation and inferential abilities in humans (Sect. 3.4.4.3). Let us see some of these implementations.

3.4.4.1 Category Theory for Ontology Building

Healy [198] describes knowledge as "a category representing an ontology formalized in a modular hierarchy of closed knowledge fragments, or theories...Expressing an ontology as a mathematical category of theories allows the machinery of category theory to be harnessed in understanding the semantics of adaptive, distributed systems".

Category theory is used here as a mathematical language for ontology building. The rationale is as follows: (i) concepts form a structure, so it is possible to build the category of concepts and (ii) structures can be mapped onto each other, so we can find a functor that preserves the structure of two different category concepts. Thus, a model is a functor where the domain is the classifying category and the codomain any category with a similar structure.

This categorical approach to ontology building, consists of defining the category of concepts **Concept**. In Fig. 3.21 it is shown how the concept T0 which is a geometric theory, can be used for forming an ontology within the category **Concept**. Morphisms in **Concept** relate theories, for example $s : T0 \to T1$, expresses the partial association of the syntax of a concept $T0$ with concept $T1$. Colimits are used in the ontology in order to provide greater specificity, as Fig. 3.22 shows. The dual of colimits – limits – provide the opposite direction, that of abstraction. But so far, only syntactic properties of theories expressed in predicate calculus have been tackled. In [17], Healy et al. go further, and in the second part of the paper, they provide a mathematical semantic model, in which these theories are grounded to network components. Thus, the syntax of the geometric theories in the example above is now given in terms of the architectural structure of neural networks.

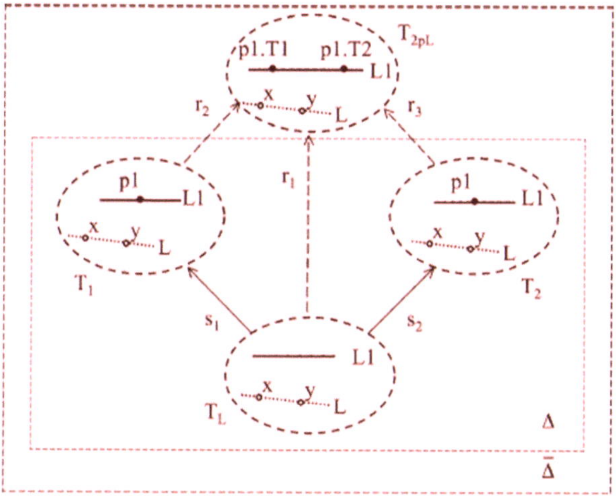

Fig. 3.22 The figure depicts the colimit T_2pL of theories T_1,T_2, and T_L. The colimit theory glues the theories T_1 and T_2 along their common sub-theory T_L (Figure from [17])

A network architecture A, formed of a set of neurons, together with an array of the connection weight values w of that set is modeled as the category $N_{A,w}$. An object of $N_{A,w}$ is defined by the pair (p_i, w), where $p_i = 1, 2, \ldots n$ represents the nodes of A and w represents the set of output values for p_i connections. A morphism $m : (p_i, w) \rightarrow (p_j, w')$ of $N_{A,w}$ is defined by a set of connection paths (synapses) between the set of nodes (neurons) and their weight states, (p_i, w), and (p_j, w').

Functors now transport the invariant structure across the category **Concept** and the category $N_{A,w}$, M : **Concept** $\rightarrow N_{A,w}$. The main idea is that learning can be modeled as a transition between categories. A functor is used to model the structure-preserving associations between categories. On the other hand, colimits express the learning of more complex concepts through the re-use of simpler concepts already represented in the connection weight memory of a neural network. The categorical approach to concepts makes it possible to model the declarative semantics of systems, such as neural networks, which are distributed over many components.

3.4.4.2 Category Theory in Semiotics

In [199], Neuman and Nave address two major problems in modeling concept formation: family resemblance and contextuality, borrowing concepts from category theory, in particular pushout and its dual, pullback. The formal theory for concept formation is based on two steps, first there is a bottom-up process that is modeled as a pushout, and then a top-down process which is modeled as a pullback. Let us see how the authors explain the concept formation of "Dog" in a child. Sign-mediated concept formation starts in the child when the mother assigns the same sign to at least two different objects. For example, the mother may refer to a Great Dane and a Chihuahua as "Dogs". We can express this with the following diagram.

Next, a third object, for example, a German Shepherd, is introduced to the child by the mother. The resulting diagram tells us that a Chihuahua is like a German Shepherd, and a Great Dane is like German Shepherd. These are local similarities, but there is an obvious global property, the three objects are all dogs, which cannot be inferred directly from the diagram bellow.

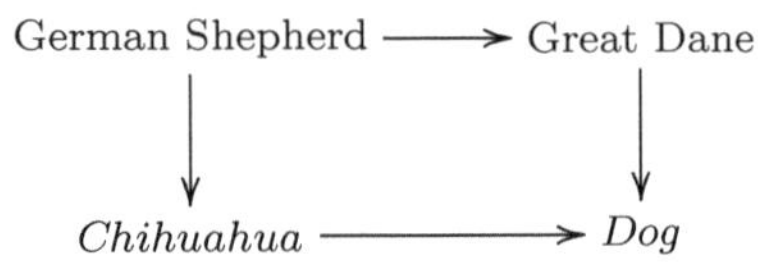

In order to model the global property of family resemblance we need to define Dog*, which may be constructed through a pullback. The dual top down process is modeled with the following pushout diagram.

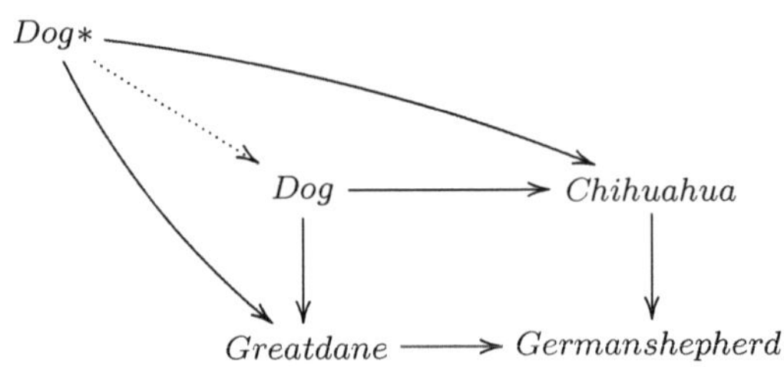

The same authors have applied category theory to understand the mechanisms of coding that living systems posses, and in particular genetic coding [18]. This work may be labeled as biosemiotics, which is the field of theoretical biology that studies the nature of coding and communication of signs in living organisms. The informational approach followed in Neuman's work, is different from the engineering approach based on Shannon's theory. The divorce between information and meaning in the engineering view, is clearly expressed in Shannon's words: "Frequently the messages have meaning; that is they refer to or are correlated according to some system with certain physical or conceptual entities. These semantic aspects of communication are irrelevant to the engineering problem [200]". For Neuman, biological systems are not information-processing systems but "meaning-making systems". Neuman defines meaning as "the effect that is produced via semiosis through interaction between at least two parties". This approach is sympathetic with the view defended in this book in Chap. 7, that states that brain coding is of form and meaning.

One of the dogmas in biology avows that the flow of information is unidirectional i.e., from DNA triplets to RNA triplets to amino acids. However, we know today that the dogma is not entirely true. It has been found a "reverse transcription" flow

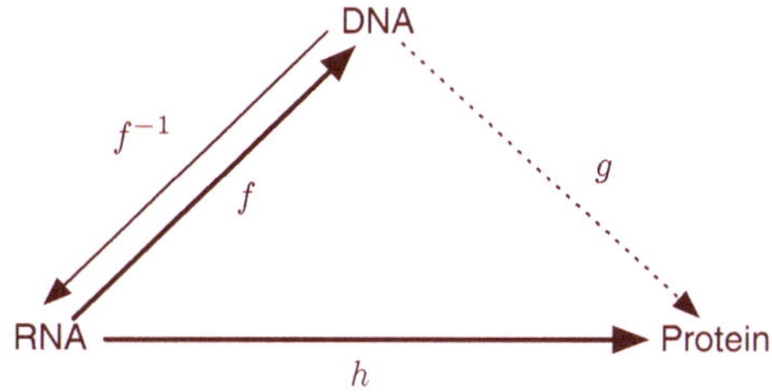

Fig. 3.23 Information flow in the genetic system modeled after [18]. The flow of information from DNA to RNA and protein is represented by the arrows f^{-1} and g. Thus, given the arrows f and h we can determine f^{-1} and g

of information, from RNA to DNA that could be at the basis of the mechanism by which, for example, RNA tumor viruses infect normal cells and turn them into cancer cells. This is however, not surprising if we reckon that biological functionality is multilevel and that there is not any priviledge level of explanation e.g., the genome. The real challenge in biological science is to determine at which level and under which principles physiological function is integrated. In addition, it has been hypothesized [201], that under certain experimental conditions, DNA can be translated into protein without the mediation of RNA, that is, a translation from DNA to amino acids, which in categorical terms is the composite of f^{-1} or reverse transcription and h or translation via the commutative Fig. 3.23.

The paper produces a mathematical solution using category theory to the following problem: how systems translate information encapsulated in DNA into proteins along two different paths: (1) from RNA to DNA; and (2) from RNA to protein. The mathematical approach to this problem relies on the concept of isomorphism. By the definition of isomorphism given in Definition 3.18, we know that a map f is an isomorphism if there is exactly one map g such that $h = g \circ f$ where $g = h \circ f^{-1}$ For a genetic system, it is possible to establish an isomorphism between the four DNA bases and the four RNA bases which is represented in Fig. 3.23 with the arrows f and f^{-1}. Moreover, this allows to reformulate in mathematical terms the question of how information flows in the genetic system. In this view, the DNA does not contain information in the sense given by Shannon, rather it has value. The value is "the effect, mainly the generation of protein, through the symbolic mediation of RNA, which constrains the different possible interpretations of the DNA for self-replication".

3.4.4.3 Category Theory in Cognitive Psychology

There are psychological categories, that at first sight, do not have too much in common with the mathematical categories that this book describes. In "To cognize is to categorize revisited: Category theory is where Mathematics meets Biology [202]", it is briefly sketched the possible parallelism between both kinds of categories. Let us look at this with more detail now. On the one hand, a mathematical category is an algebra of mappings among things that can be composed and where composition is associative. On the other hand, in psychology, categorization is referred to as the process by which distinct entities are treated as equivalent. The output of the process of categorization are categories, which are concepts that we necessarily build in order to cope with reality. Within this view, category and concept are assumed to have identical meaning. Mental concepts or categories are created based on the *principle of similarity*: an object falls under a concept if we can determine that the object is sufficiently similar to the other category members. Thus, an instance i falls under a concept or category C *iff* $\forall feature f \in C, f(i)$is true. This theory of concepts as categories of similar items has been predominant in theory of mind since the seminal work of Rosch and Mervis [203] and Rosch [204]. For a detailed description of this view and its criticisms see [205].

William C. Hoffman [181], shows that the basic structure is the same for both kinds of categories. He establishes a correspondence between psychological and mathematical categories based on equivalence. Such an association allows us to set the basis for a mathematical investigation of cognition, perception and emotion, which are the fundamental components for consciousness. The mathematical theory of categories makes available tools for the deduction of hypothesis for structure and function in neuropsychology. The intrinsic categorical nature of the mind is clearly stated in this quote:

> the computation involved in basic human affairs is primarily topological or categorical rather than computational in the traditional computer-science sense of that term. Every percept is a geometric object in the mathematical sense. And concepts and emotions, to be sufficiently general, must be categorical.

Hoffman uses the category of sets **Set** and the category of topological spaces **Top**, the former models "trains of thought" and the latter is able to cope with percepts, concepts and emotions that occupy the mind.

Philips et al. [182], use category theory to show that, human inferential abilities like transitive inference and class inclusion, involve the dual category theory concepts product and coproduct. Children around 5 years of age develop what is called *Transitive Inference* which is for example, given that Bertrand is taller than Ludwig, and Ludwig is taller than John, one may infer that Bertrand is also taller than John. Class inclusion develops later in children and consists of the ability to discern between the cardinality of classes and subclasses, for example there are more animals than cows. Category theory shows that these abilities can be formally connected. Transitive inference can be modeled with the product, and class inclusion with its dual, the coproduct. This fact would explain that these two reasoning abilities have "similar profiles of development", because they involve related sorts of processes, namely product and coproduct

In a later paper, [206], Philips and Wilson use category theory to explain systematicity in human cognition. Systematicity is defined as "the property of human cognition whereby cognitive capacity comes in groups of related behaviours". In this work, systematicity is perceived as a necessary consequence of a higher-order theory of cognitive architecture, this would make either the symbolist and the connectionist approaches inadequate to explain systematicity. While connectionist and symbolist theories are both first-order, category theory, and in particular adjunctions, allow us to deal with the higher order structure, which is more appropriate for cognitive functions modeling.

Definition 3.28. An adjunction consists of a pair of functors, $F : C \to D$, $G : D \to C$, and a natural transformation $\eta : I_C \to (G \circ F)$, such that for each C-object X and C-arrow $f : X \to G(Y)$, there is a unique D-arrow $f^{\#} : F(X) \to Y$, such that the following triangle commutes:

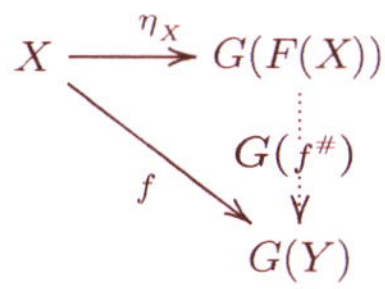

An adjunction is a natural transformation between functors, natural here means particular realization of the functors. By analogy, systematicity in Philips and Wilson words "is natural in that it does not depend on a particular representational scheme. Hence, the explanation does not depend on ad hoc assumptions about internal representations". In their conclusions, the authors suggest that a more integrative theory of cognitive architecture could be attained by including acquisition of representations which is not tackled in the paper. Indeed, why an object belongs to a particular class and not to another is not explained but assumed. Finally, the incorporation of category theory into the Bayesian theory is stated as a promising future line of research. Undoubtedly the creation of a category of Bayesian models will need of ingenious mathematicians.

Chapter 4
Elementary Principles in Cognitive Systems Modeling

4.1 Introduction

This book is about thinking, in all its ontological, epistemological and operational facets. It deals not exclusively with the process of thinking, but also with conditions that must be created and laws that must be fulfilled, in order to think about the process of thinking. Arguably, human cognition may be largely a emergent phenomenon. How cognitive functions "emerge" from the nested hierarchical architecture of neural systems, is a question that has been posed with a myriad of approaches. Concepts that traditionally belong to the philosophy of science like emergence, reductionism or principle, have started to receive some attention from researchers in cognitive and neural science [207–210].

In this Chap. 1 I propose a framework for modeling in cognitive science. Section 4.2 explores the concept of reductionism. I will define reductionism and I will show that, for the purpose of modeling complex systems, the reductionist approach can only give us a fragmentary picture of the system. Section 4.3 discusses formalisation and its limitations. Emergence is introduced in Sect. 4.4, this concept is explored in a rigorous way, trying to provide a formal description of emergent properties within a context relevant for cognitive scientists. At last, in Sect. 4.5 three principles for modeling in cognitive systems – locality, hierarchy and multiplicity – are described. It ought to be remarked that in this chapter, modeling is not seen as quantitative prediction data, that kind of models were described in Chap. 2. Here, I present a survey of modeling in a more basic or fundamental way. Modelers should not be satisfied to merely demonstrate a more or less optimal fit to data, the goal of modeling is to increase our understanding of systems [211].

J. Gómez-Ramirez, *A New Foundation for Representation in Cognitive and Brain Science*, Springer Series in Cognitive and Neural Systems 7, DOI 10.1007/978-94-007-7738-5_4,

4.2 On Reductionism

Reductionism is a term that has gained considerable bad press within certain *cultural milieux*, notably in the social sciences where reductionism is usually portrayed a faulty and dangerous way of studying complex phenomena using simplified and overarching mathematical generalizations. Admittedly, reductionist statements ornamented with some obscure mathematical terminology by some, has served to brutalize social reality, and minimize environmental influences for the most self-serving reasons. As an example of this, there is a very colourful dispute, published in the journal *Mathematical Intelligencer*, between the mathematician and number theorist, Neal Noblitz and Herbert Simon, Nobel prize in Economy [212]. At the origin of the argument was the politologist and Simon's friend, Samuel P. Huntington, the same that wrote later on, in 1993, The Clash of Civilizations, an essay that was very influential in G.W. Bush's State Department.

In a 1968 article [213], Hungtinton's views on societal issues supported by mathematical equations caused the irritation of some sensitive readers with mathematical background. Hungtinton defined the following equations in order to justify his conclusions about the impact of modernization in developing nations:

$$\frac{SM}{ED} = SF, \frac{SF}{MO} = PP, \frac{PP}{PT} = PI$$

where SM is *social mobilization*, ED is *economic development*, SF is *social frustration*, MO is *mobility opportunities*, PP is *political participation*, PT is *political institutionalization* and PI is *political instability*.

The problem is that Huntington never formally defined what the variables used in those formulae, *social mobilization*, *political instability*, exactly mean. He neither provided the reader elementary facts like the units in which the variables must be measured or their range of possible values. Hungtinton's conclusions, supported by trivial and badly formulated mathematics, led to surprising consequences. For example, Philippines was a "stable" country and France was not, or South Africa, during the apartheid was a "satisfied society" [214]. If we acknowledge the big mismatch between the mathematical equations and the social and economical facts that the formulae pretend to explain, the conclusions are not surprising at all, they are essentially the consequence of bad mathematics. This was pointed out by Neal Koblitz in "Mathematics as Propaganda" [215]. Despite the evident bad formalisation practised by Huntington, Herbert Simon came to the politologist's defense with a paper on "Some Trivial But Useful Mathematics" (Simon, 1987, Some trivial but useful mathematics, unpublished paper). A vehement exchange continued and other mathematicians like Saunders Mac Lane and Serge Lang, joined Koblitz against Hungtinton's simplistic reductionism [216].

Reductionism and mathematization, are a danger only when used with spurious interests and limited knowledge of the mathematical structures that are introduced in the explanations. Hungtinton's case is a conspicuous example of this, alas, it is not an unique case. The psychoanalyst Jacques Lacan, suggested, and convinced some people with no mathematical background that Don Juan was a compact topological space [217]. Having shown the dangers of reductionism and mathematization in social sciences, we must not misunderstand them. In social sciences and in cognitive science too, the ambiguous and strictly verbal account of the facts still prevail. There seems to be some sort of irrational or pathological fear of reductionism and mathematization within the realm of social science and cognitive psychology. The following unattributed quote expresses this superbly

> If in the physical sciences, one is able to successfully reduce complex phenomena to a simple rule or model, one is awarded the Nobel prize, the reward for a similar attempt in the social sciences is to be pilloried in the New York Review of Books.

With this in mind, we can readily see the necessity of reconfiguring the concepts of reductionism and formalisation on more rigorous ground.

4.2.1 What Is Reductionism?

A reductionist approach explains a phenomenon C, through a simpler one, S. It is by that account that reductionism gives reductive explanations. However, it might be reminded that there is no understanding without reduction. In philosophy of science, to reduce is commonly understood as to express laws and rules of secondary sciences in terms of primary ones [218] i.e., the secondary science of thermodynamics was explained using concepts and laws belonging to Newtonian mechanics, the primary science. In mathematical logic, to reduce is considered as equivalent to deduce. Thus, given a set of premises $\{P\}$ we deduce a new statement $\{s\}$, however, $\{s\}$ is not really new because it was already contained in P.

We must distinguish between different kinds of reductionism e.g., higher vs. lower and partial vs. total reductionism. Bunge [219] establishes a differentiation between higher reductionism and lower reductionism. Indeed, the etymology of to reduce is *reducere*, that is, to lead back and not necessarily toward smaller scales. One example of higher reductionism is to state that a calculator is intelligent because it is fabricated by an intelligent person. An example of lower reductionism is to explain the intelligence of an animal as a set of automatic mechanisms. On this view, emergent phenomena can be more or less complex than the elements that give rise to them [220]. Accordingly, *Behaviorism* could be seen as lower reductionism – a complex mental function e.g., emotion, may be reduced to a set of preconfigured states e.g., the states of an automaton which are much simpler than the mental functions. Both approaches are reductionistic, but it is important to note that the higher reductionism is, epistemologically speaking, more untenable than the lower one, because it leads to obscure explanations about not necessarily complex phenomena. In other words, it explains something by means of something else even more complicated *ignoti per ignotum*, as in the above example: a calculator is reduced to something more complex, a person. The lower reductionism, on the other hand, constitutes the very basis of the modern science, which roughly consists on the study of any object by reducing it to its elementary components and properties. Particle mechanics is deducible from quantum mechanics and solid rigid kinematics is deducible from solid rigid dynamics. Thus, particle mechanics and solid rigid kinematics are partially reducible from quantum mechanics and solid rigid dynamics respectively. Theories that are totally reducible are isomorphic [221]. Attacks on mathematical-reductionistic approaches on psychology miss for the most part the point that reductionism is two-sided. Psychology is obviously not totally reducible to biology but it can indeed be partially reduced. In short, let $\{\text{Psy}\}$ and $\{\text{Bio}\}$ be two theories, and let $\{\text{A}\}$ a set of assumptions not included in either $\{\text{Psy}\}$ or $\{\text{Bio}\}$. Then $\{\text{Psy}\}$ is fully reducible to $\{\text{Bio}\}$, iff the predicates of $\{\text{Psy}\}$ follows logically from those of $\{\text{Bio}\}$. On the other hand, $\{\text{Psy}\}$ is partially reducible to $\{\text{Bio}\}$ iff $\{\text{Psy}\}$ follows logically from $\{Bio\} \cup \{A\}$ and not from $\{\text{Bio}\}$ alone.

Reductionism must reckon with two main challenges. One is of ontological nature or what are the elementary components belonging to the complex studied object? and the other is epistemic or how well the lower components of a complex object reflects its nature and essential properties? These two problems are indeed almost perennial matters in the history of philosophy and science. Therefore no magic recipe or simplistic formula can be expected here, rather there is a need for a theory, within a general framework, where the questions exposed above can be conveniently treated. In order to fully understand a complex phenomenon is by no means enough to decompose it in its "atomic" components. Atomism, as a matter of fact, subscribes a strong bottom-up causation placed in a temporal framework where time is unidirectional. In complex systems, for example, biological ones, not only the physical factors have to be considered, ecological factors are necessary for the obtention of powerful explanations. Even though the environmental or cultural aspects are, of course, ultimately sustained in physical causes, such causes could be inaccessible or irrelevant. Causality is not always exclusively unidirectional, there is also a downward causal flow. We are very close in science to reduce as much as we can, the next step must be a work on synthesis. A full understanding of complex systems will come only form integrating the reductionism methodology into a more all-inclusive approach [222, 223].

4.3 On Formalisation

Descartes' invention of analytical geometry supposed an astounding conceptual revolution. The universal order, with its objects and laws, could be mapped and studied within the Cartesian concept of space. This is conventional knowledge, however, we should not take for granted the dramatic change that the inclusion of algebra in geometry supposed at that time. Geometric forms which in the pre Descartes era were thought of as synthetic mental ideas, could be translated into algebraic equations. This paved the way, three centuries later, with the work of

Lobachevski and Bolyai, to non-Euclidean geometry and its application in quantum physics [224]. However, the Cartesian axis seems a very feeble skeleton to sustain the whole body of the universe laws. In words of David Bohm [225], reality is a multi-dimensional reality and only under very concrete conditions can be satisfactorily simplified to three dimensional reality. Reckoning this as true, it exists the danger to misunderstand ontology with epistemology, or said plainly between, on the one hand, the things that populate the world, and on the other the methods to know those things. The view defended in this book is that formal systems, expressed by means of diagrams or equations, can be an extraordinarily efficient and successful mechanisms to explain phenomena and represent features of the world, including those that represent the world. The point to be retained is that formalisations are methodological tools and not ontological simplifications. For example, a map of London is not the same as the city of London, however, the map is a formal representation of a physical system, or to put it a different way, there is a homomorphism that preserves significant aspects of the system that is being modeled and the model itself. The fact that the map captures the structure existing in the city is evident in this. Note that the relationship between the city of London and a London's map is homomorphic rather than isomorphic. The map has to be simpler than the world it models. This means that every so often the map may be wrong or is insufficient.

When we succeed in representing an external object of the real world in our head, a map, or to be more precise a homomorphism that preserves the common structure of both the external object and the mental object, has been established. Using the theory of categories, if we specify the category M of mental objects and their homomorphisms and the category E of external objects and their homomorphisms, a functor $F : M \rightarrow E$ constitutes a model of the external world. Let P a physical system, given by a domain P_D and a set of relationships P_R, and let M a mathematical system, given by a domain M_D and a set of relationships M_R. A homomorphism h, is a mapping from P_R to M_R that preserves the structure in an appropriate way [214]. The psychophysical homomorphism h constitutes a small scale reproduction of the natural laws that rule the physical world. In the words of Dehaene, "we carry in ourselves a universe of mental objects whose laws imitate those of physics and geometry" [226].

When moving from the *hard sciences* e.g., physics, to less hard sciences e.g., psychology or sociology, the formalist credo does not see as well received as it is in physics. Needless to say, putting our daily experience and intuitive concepts into formal terms is not an easy task by any means. However, the formalist endeavor in science – hard or soft – is unstoppable, and as formal descriptions of the phenomena studied progress, ordinary accounts are abandoned. Clearly, this replacement of the ordinary for the mathematical-formal is a constant in the books of the history of science. Quoting R.D. Luce:

> and as honest scientists, as far as we disentangle the structure we need to begin to describe it in formal terms [227]

4.3.1 The Limitations of Formalisation

Next, I sketch six general obstacles to formalisation. In what it follows I hope to shed some light in a crucial point: formalisation has not only an epistemological dimension that can be understood as a method to gain some knowledge about the world, but also an ontological dimension, formalisation is an inescapable cognitive activity intrinsically related to perception.

4.3.1.1 Definitions in the Dictionary Are Circular

When asked by a child, for example, about the definition of one word, we easily realize that very simple questions may require long and complicated answers. Let us obviate here all the difficulties rooted in the process of categorisation of everyday concepts and in its formulation in natural language and focus in the definitions of the concepts. The definitions that one finds in the dictionary are sometimes circular and ambiguous. In natural language, a circular definition is a description of the meaning of a word that assumes a prior understanding of the term being defined. For example:

fast -adj. : *swift*; *quick*; *speedy*

swift -adj. : *fast*

Or another comical and real example is:

see -verb : See *see*

The basic component in a dictionary are the lexemes (words) which as it is shown above, are sometimes ill-defined with circular definitions. In mathematics is possible to start with more basic concepts than lexemes. The indecomposable elements in mathematics are objects that can be seen as elementary particles with a minimal amount of ambiguity in their definition. The indecomposable elements hold the Indecomposability Principle established by Brouwer that states that the continuum cannot be partitioned into two nonempty pieces [228]. The idea I am defending is simply to promote the use of formal languages e.g., mathematics rather than natural languages e.g., English, in cognitive and brain science. A good reason for taking this strand is that the basic concepts of formal languages are infinitely closer to indecomposability than natural languages [217]. It must be emphasized that this approach is radically different from the language of thought hypothesis (LOT), proposed by the philosopher Jerry Fodor, that postulates that both thought and thinking coexist in a mental language [229]. LOT is linguistically biased – thinking is seen as the product of syntactic operations defined over representations. The problem with LOT is that as it exclusively focuses in linguistics, and obviates other languages like mathematics, which is an extremely efficient source of new concepts and tools to understand reality (thought included).

4.3.1.2 Obstacles to Formalisation: The Plurality of Definitions

Formal logic definitions are not always enough to fully describe a mathematical concept. Indeed, in textbooks of mathematics, the definitions of new concepts are self-included in definitions but they usually need of additional lemmas and examples in order to be properly defined. It ought to be noted that in mathematics, unique definitions for mathematical concepts do not exist, not even for those as important as function or structure. This is because these are concepts used in several areas of mathematics in many different ways, so no single definition of function has been universally adopted. For example, the concept of morphism which recurs in much of contemporary mathematics and in particular in this book, has different denotations. For example, in set theory, morphisms are functions, in linear algebra linear transformations, in group theory group homomorphisms; and in topology continuous functions. This lack of unity in definitions, even for some of the most fundamental concepts, ought not to drive us to wrong conclusions about the useless or impossibility of the formalist program. Au contraire, the obtention of formal definitions provides a common setting where everyone knows what the others are intended to say, even if restricted to some particular domain, situation or context.

4.3.1.3 Obstacles to Formalisation: The Sensitive Coding

Another problem alleged against formalisation is the sensitive coding (here the term coding is borrowed from informatics and it is intended to be equivalent to formalisation). Obviously, when encoding any process or set of data, some relevant parts of information can be lost. This occurs because the encoding does not capture the invariance of the object that is being coded, in consequence the structure is not preserved through the codification process. The coding process can be seen as a morphism, in order to capture the invariance such morphism should be an isomorphism.

Spikes or action potentials are measurement values that provide information about the state of the system. Thus, using Shannon's communication scheme [230], a neural code is given by an input set S which represents stimuli, an output set P which represents spike patterns and a coding function f_c that maps S into P, $f_{code} : S \rightarrow P$. Neural decoding is the inverse process, f_e operates in the P space to produce an estimation of the stimulus. $f_e : P \rightarrow S$ Shannon's theory of information is not a mainstream technique in neuroscience, and no reference at all can be found in major neuroscience textbooks like Kandel's The Principles of Neural Science [231] or Andersen's The Hippocampus Book [232], while in Squire's Fundamental Neuroscience, information theory is only mentioned in the last section of the book in relation with theories of consciousness [233]. The limited use of information theory in neurobiology might be found in the failure to provide a quantitative definition of neural information. In the 1950s and 1960s neuroscientists and cyberneticians warned about the difficulties in applying the concepts of information theory to the nervous system. To make this point more clear, it is worthy to recall statements from scholars like George Moore "at best the concepts of information theory are metaphors when applied to the nervous system [234]". Nevertheless, in the last decade, information theory has regained the attention of neuroscientists. Information theory concepts are principal in the understanding of how information carried in the form of neural activity patterns is precisely encoded in the brain [235]. In order to understand how higher cognitive functions like memory or perception work, we need to know how information is encoded by neurons and populations of neurons [236]. Information theory provides a theoretical framework in which it is possible to quantify the amounts of information carried by neural activity. The important point here is that as different codes may describe the same neural activity, we need to be able to compare how much information is hold by these codes.

4.3.1.4 Obstacles to Formalisation: The Gödel's Incompleteness Theorem

The first and second Gödel's incompleteness theorems are used for some scholars as the kill all argument against formalisation [237]. Gödel's first incompleteness theorem says that any axiomatisable theory that is adequate for arithmetic is incomplete. This, some argue, would imply that any attempt to capture some arithmetical aspect of the world using a formal system, will not be able to prove all the truths about that aspect of the world. In the same vein, Skolem's paradox shows that no formal axiomatised system can capture our intuitive notion of set [238]. This is, however, far from surprising especially if we are aware of the simple fact that the formal systems that we generally construct cannot completely capture the aspect of reality that they describe. This is analogous to the incompleteness of arithmetic since our formal axiomatisation of arithmetic cannot capture everything about the mathematical system itself.

One main lesson to get from Gödel's incompleteness theorem[1] is that the foundational relationship between mathematics and mathematical logic is ill-founded. Indeed, in order to verify the correctness of a theory T, it is necessary to build a stronger theory T'. This leads to an infinite regression because the completeness of T requires a new T' such that $T' \supset T$, which needs of a new T'' such that $T'' \supset T'$ and so on. The Gödel's theorem demonstrates the impossibility of establishing the non contradiction of arithmetic solely by methods borrowed from arithmetic or weaker theories [239]. The Gödel's deadlock means that every time we build a formal system, in order to test its completeness, one stronger formal system has to be additionally built. As there is not a finite method to guarantee the completeness of any formal system, the gap between formalisation of a system and the system itself

[1] *First Incompleteness Theorem*: Any adequate axiomatizable theory is incomplete. In particular the sentence "This sentence is not provable" is true but not provable in the theory. *Second Incompleteness Theorem*: In any consistent axiomatizable theory (axiomatizable means the axioms can be computably generated) which can encode sequences of numbers (and thus the syntactic notions of "formula", "sentence", "proof") the consistency of the system is not provable in the system.

seems abyssal. But this is not any impediment at all for the formalist programme. For example, in either Set Theory and in Computer Programming, formal definitions that exemplify this infinite regression abound. The axioms of Peano for the natural numbers is a good example of this. The formal definition of the natural numbers, one of the most basic structures, is recursively defined. In addition, it is worth noting that the concept of formalism proposed in this book is wider and more general than that is used in the logical tradition that focuses on formal consistency and completeness, which are purely syntactic concepts. Formalisation can be seen as the cognitive process that necessarily mediates in the grasping of the regularities existing in the world, and formalism as the output of such process, bringing together the mathematical entities that reflect the structural invariance between the real system and its formalisation.

4.3.1.5 Obstacles to Formalisation: Formalisation as Opposed to Intuition

It is necessary to abandon the platonic idea of mathematical forms and theories as an extramundus thing. Gödel himself was a champion of Platonism in mathematics, not in geometry which he conceived as physically grounded in the world, but in Set theory. In this view, the objects of Set Theory are Platonic forms because they are non-physical yet objective. Intuition makes possible to perceive truths about these objects with which we have no physical interaction [240]. Let us for a moment assume the validity of the Gödel's platonism credo for Set Theory. The Platonic forms to which we have access only through intuition are committed to the law or order that we perceive in nature, it follows then that those intuitions must be governed by rules as well. Both intuition and set-theoretic objects are mutually constrained by the morphological laws or legality using Jean Petitot's notation [241].

It is vulgarly assumed that intuition covers all that is not or can not formalised. At this point is convenient to shed new light on the concept of intuition. Intuition and formalisation are not on opposing sides, rather intuition is at the core of formalisation because intuition naturally converges into formalisation. There is a tendency to understand formalisation as a process that progressively limits the field of the intuition which we can be defined as that that is not (yet) formalized. This is a dangerous myth that must be debunked. First, even in the case that formalisation was *une chôse pour des elites*, this elite should be listened to, at least, as much as those lacking the skills that the formalists posses. Second, there is no such a thing as unstructured cognitive state. Any mental content, in order to be attained, needs of the intermediation of perceptual schemas (forms) that configures the agent's perception [242]. Mistakenly, formalisation is seen as a prerogative of a small elite composed by those that find themselves comfortable in the mathematical and logical arena, in opposition to intuition, which is seen as a natural gift that everyone has, and therefore free of pedigree nor specific skills. It might be remarked here, that all explanations or descriptions are rooted in some context or theory. The definition of any object makes sense, only as far as is related to a theory, which must be formalised (mathematised) to be operational.

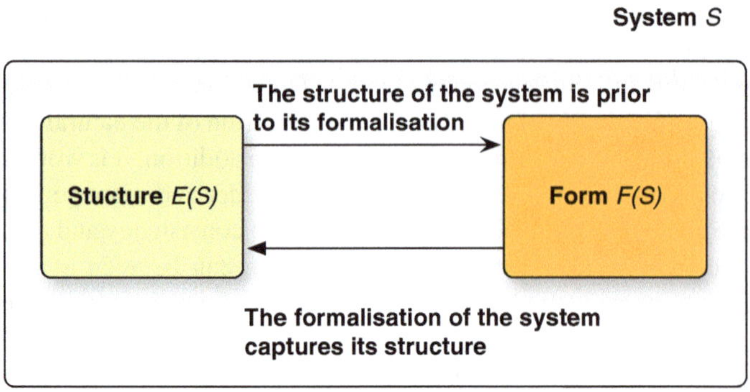

Fig. 4.1 The structure is prior to its form and the formalisation captures the structure of the system

The cognitive agent is always mapping the outer repeating patterns into its own inner structures. However artificial it may seem, formalisation resides naturally and primitively in the cognitive agent i.e., formalisation supersedes intuition. The world out there is nomologically organised, and the only option for us, as cognitive agents, is to interact with it. The reason for this is as simple as to admit that conceptualization is theory based. Mathematics or any other formalism, becomes relevant to science when the scientist is able to disentangle the structure in the observed phenomena. The symbols that the formalism is made of are necessary to keep track of those structures captured in the scientist's head. The scientific study of cognition will, of necessity, require a formal language able to seize and to express in an systemic and comprehensible way the structural invariance residing in the phenomenon of cognition [243–245] (Fig. 4.1).

4.4 Emergence on Systems Modeling

The modeling cycle in science necessarily begins with a reductionist approach to create the simplest possible model. The modeling process generates an understanding of the underlying structures, as components are represented with mathematical objects. Phenomenological descriptions cannot be captured by hypothesis that solely rely on the components' properties and minimal models might grow in complexity. The informal notion of "complexity" has a number of approaches and definitions. For example, the Kolmogorov complexity of a given object is the amount of information required to describe the object. Here we are interested in organizational complexity. The organizational complexity, in order to be studied, needs a scale hierarchy representation of the system. In this view, larger scale systems may impose boundary conditions upon the faster dynamics of lower scale systems [246, 247].

Emergence may be seen as a manifestation of complexity. However, emergent properties do not necessarily arise in complex systems. There are many examples of emergent phenomena, some of the most common are transitions between solid, liquid and gaseous states, honeycombs, life, markets, economies, cities .... As this is a book about cognition, we will explore the connection between emergency and complexity. The mind can be studied as an emergent phenomenon that arise, in last term, from purely physical processes that occur in the brain which is a system with a tremendous organizational complexity.

4.4.1 A Few Notes on Complex Systems

In order to be effective in complex system modeling, we cannot rely on disquisitions about how complex a system is by classifying properties as global or local. The main problem with this procedure is that it assumes three things that cannot be taken for granted:

- The components of the system have clear cut borders
- The frontier between the system and the environment is a clear cut border
- The properties of the system are always observable and measurable

Perhaps, for a better understanding of complex systems we must outline the work of system theorists. Ludwig von Bertalanffy, father of general systems theory, defines a system as a "set of elements standing in interrelations". Mario Bunge has accomplished a colossal work, writing a veritable opera magna in eight volumes that goes from Ontology to Sociology [221]. Bunge, leaded a formidable project[2] called Exact Metaphysics. In essence, Bunge is a systemist, the motto "everything in the universe is a system or a part of a system" and "every system has global properties that its components lack", has been formally defined in Bunge's work with some recent variations and adjustments as we will see next.

Given a system S, a model of the system, $m(S)$, can be expressed as follows:

$$m(S) =< C(S), E(S), S(S) >, \tag{4.1}$$

where $C(S)$ is the set of components of system S, $E(S)$ is the set of components of the environment relevant for S, that is to say, items that act on or upon items of $C(S)$, and $S(S)$ is the structure of system S, which is the collection of relationships between the parts of $C(S)$ and between $C(S)$ and $E(S)$. Moreover, $S(S) = S_I(S)US_E(S)$, where $S_I(S)$ or endomorph structure represents the internal structure of system S, and $S_E(S)$ or exomorph structure is the structure created by the relationships between internal items in $C(S)$ and their relevant neighbors in $E(S)$.

---

[2]This project has not been completely abandoned. Indeed the Society for the exact philosophy organises every year a congress

In [220] Bunge included the mechanism $M(S)$ to the above model.

$$m(S) =< C(S), E(S), S(S), M(S) > \tag{4.2}$$

However, the addition of $M(S)$ does not resolve one of the main objections of this model: the distinction between the system and the environment is too clear cut. It is well known that one of the main features of a complex system are the indeterminate nature of its boundaries. The decision about where the system finishes and where the environment begins is ultimately made by the observer which is not included in the model.

Nils Baas has developed a model that deals with emergent properties in complex system including the observer [248]. The definition of emergence rests on the existence of an observer with its own observational mechanisms. Thus, one property is emergent when it cannot be observed in one system S_1 but it may be positively observed in the resulting system S_2, which is the new structure generated through the interactions of the components inside S_1. Then,

$$S_2 = R(S_1, Obs^1, Int^1) \tag{4.3}$$

where S_1 is the initial system, Obs^1 are the observational mechanisms and Int^1 are the interactions between neighbor components of S_1. Therefore p is an emergent property iff can be observed in S_2 and not in S_1. Nevertheless, the inclusion of the observer in the model is problematic because it may lead to infertile disputes, for example, the realism/anti-realism clash in analytic philosophy. The realists subscribe the existence of a world of objects independent of anyone's linguistic practices or conceptual schemes. On the other hand, anti-realism, states the predominance of the language over the thought and of the thought over the reality [249]. According to anti-realism, it is useless to try to answer whether one sentence is true or false and this is because in actual fact, the realist and the anti-realist support rival systems of logic. Dummett, an anti-realist, suggests that a definitive resolution of such metaphysical debates can be obtained only if we give predominance to the language. Consequently, the confrontation between realists and anti-realists vanishes, because each opponent would be concerned with a particular type of language – so one might be an anti-realist about arithmetic but a realist about the past [250]. To sum up, the inclusion of the observer, poses some metaphysical complications which it would be preferable to avoid.

4.4.2 A Few Notes on Emergent Properties

The principal source, at less at philosophical level, in the study of the phenomenon of emergence, is that of the confrontation among atomism and holism. The common attitude toward atomism reflects a misleading conception: atomicity is not a property that an object possesses, rather is a model or category good at describing objects at some particular scale. On the other hand, holism is usually referred by the motto "the whole is more than the parts". Holism states that in a given system, there are properties of the system that cannot be determined or explained by the system's components alone. In this view, the properties of a system as a whole that cannot be found in any of its individual components are called emergent properties [251]. For example, the GNP of a country is an emergent property i.e., it is possessed by the whole ensemble of citizens and not by any single individual. It is necessary to clarify two points here. One is that emergence is an universal phenomenon present in a number of systems, therefore, it is simply a bad idea to study it with esoteric mereologies or as a mysterious phenomena that requires even more inexplicable assumptions [252]. The other point that deserves attention is that it is possible to analyse emergence with rigour and precision using mathematical tools. More generally, when definitions or explanatory theories of complex systems are required, purely verbal approaches are in most cases unconvincing. For example, the system composed by two atoms of Hydrogen and one atom of Oxygen have properties that lack the same single atom of Oxygen. Contrarily to properties such as the melting point, weight or density which are present in either the molecule of water and in the Hydrogen and Oxygen, there are other properties, like for example, viscosity, that only acquire sense when are used in the molecule of water and not in its individual components. To call to those properties emergent or just systemic properties is from my point of view, just a matter of choice.

Once the idea of emergence as a hidden or mysterious phenomenon has been rejected, there is no objection to define that concept formally. Given a system S, a subsystem s_i of S is emergent iff s_i possesses some property that is uniquely possessed by the subsystem s_i inside the system S. Formally,

$$S = \bigcup_{i=1}^{n} s_i, \exists p | p \in s_i \wedge p \in s_j \, j \neq i \tag{4.4}$$

One important point that we can glean form the above remarks is that in order to build a model m of a given system s, first we have to identify the objects that compound the system, $C(S)$, then the objects that are not the system but affect or are affected by the system $E(S)$, and at last the relationships between the objects configure the structure of the system $T(S)$. Once all its components and their relationships have been dissected, the study of the properties of the system rests just a matter of observation and measuring. But when we turn back to the subject of complex systems, that is to say, systems with emergent or systemic properties, we quickly encounter problems. For example, if one builds the model of a multilevel system, for example a tissue made of cells, the modeling process is not without complications. First, the model requires a complete enumeration for both all the cells belonging to the tissue and all the cells that do not belong to the tissue, which act upon cells of the tissue. Second, for every cellular component one needs to account for all its neighbors and the relationships between them. Third, it is also required to describe all processes going on or upon all this plethora of cells. In addition to this, the tissue, of course, is a not a static object but inherently dynamic i.e., it is continuously suffering transformations. To be specific, cells undergo processes like dying, growing, moving or duplicating, so the components of the tissue may have changed in two different instants of time [253]. Clearly, systems modelling is a primary tool for the study of the structure of a system. However, this is conditional on the requirement that the system is disclosed in its components and in the relationships between them. The structure cannot be comprehended as a predefined set of pairs between components, because those components and the relationships between them are not fixed but changing with the time.

I want to oppose this view with that of structure as the relevant organisation of a system that is invariant under transformations (homomorphism). This argument will be introduced in the next chapter. The practical consequence of this approach is that we must shift the focus from the objects of a system to the relationships between the objects that remain invariant under transformation. Category theory highlights this by emphasizing the way that the components are connected by the edges rather than the components themselves. Thus, it is possible to define hierarchical categories in which the components e.g., nodes in a network may have different levels of complexity in the same category. For instance, we can define a category called cat-neuron that contains neurons, but also more or less complex classes of synchronous assemblies of neurons representing mental objects. This hierarchical approach paves the way to study emergence focusing on the structure of the things, and not in their properties.

Emergence in [184] is modelled as complexification of a category, an iterative process that may lead to the emergence new objects and links with strictly increasing complexity orders. The complexification process is used to model – according to Thom [254] – the four basic configurational transitions that a natural system must undergone i.e., birth, death, confluence and scission. Having modeled the actual configuration of a given system, for example a cell, as a category $\mathbf{C}$, it is possible to describe any change of configuration of a system via a functor $F : \mathbf{C} \to \mathbf{C}'$. Thus, the complexification process is a series of functors, F,G,H..., by which an initial category $\mathbf{C}$ evolves into successive categories $\mathbf{C}'$, $\mathbf{C}''$ ..., that is, $F : \mathbf{C} \to \mathbf{C}', G : \mathbf{C}' \to \mathbf{C}'', H : \mathbf{C}'' \to \mathbf{C}'''$ .... The mechanism responsible for the transition between categories that is here modelled with functors are called options. An option on a category is a list of changes in a system. The challenge of the complexification process is to find the category modeling the system after the changes triggered by options.

It is worth reminding that categories are universal classifiers of graphs and mappings. The complexification $\mathbf{C}'$ of $\mathbf{C}$ with respect to the option F may solve the "universal problem" of finding an universal solution to the problem of constructing a partial functor $F : \mathbf{C} \to \mathbf{C}'$ such that the objectives of the option are achieved. To summarize, the complexification process models a path of, for example, temporally indexed changes of configuration, and as a result new complex objects are created.

4.5 Three Principles for Cognitive Systems Modelling

This section attempts to relate emergence and other properties of cognitive systems like degeneracy, to three basic principles, namely locality, hierarchy and multiplicity. Establish this connection is an attempt to gain a better understanding in the design principles of cognitive systems, that may also be of help for researchers that try to build artificial systems equipped with cognitive abilities. In our exploration of the phenomenon of emergence, three principles are principal for modeling complex and cognitive systems. Namely, Principle of Locality (Sect. 4.5.1), Principle of Hierarchy (Sect. 4.5.2) and Principle of Multiplicity (Sect. 4.5.3).

4.5.1 Principle of Locality

The frame problem derives from the technique used in animated cartoons called framing, in which the currently moving parts of the cartoon are superimposed on the frame, which depicts the unchanged background of the scene. This problem was formally stated by John McCarthy and Patrick J. Hayes and refers to the difficulties found when trying to obtain a formal representation for a changing world [255]. The world is inherently dynamic and in continuous change, but not every action modifies all the possible states of the world. Indeed, there are persistent states and morphisms between states as well as. The interesting point here is to establish which morphisms are homomorphisms or invariant under transformation. It is thus required that we identify which states of the system are persistent and which are not. For example, if when driving a car one changes the gear, as a consequence of this action the torque in the car's engine will be drastically modified, but of course, the color or the number of doors of the car will not (because they are a persistent states). The frame problem quickly arises when we build a formal knowledge specification, to be, for example, implemented in a robot interacting in a real-dynamic world. A complete specification in the robot's knowledge base of the state of affairs that follows each of its actions supposes a endless coding of axioms.

Common sense theories have flourished in Research centers and Universities in the last 20 years, unsurprisingly the most active programs are around laboratories like the MIT Media Lab with luminaries like Marvin Minsky, the main referenced scholar in the vast literature of the frame problem.[3] Common sense theories are claimed as necessary because they could help in solving the frame problem [256]. On the contrary, the thesis defended here is that the lack to acknowledge the principle of locality, reveals the inadequacy of the habitual tactics based on logics or in common sense theories to deal with the frame problem. Here, no solution

[3]There is some controversy about the paternity of the Frame Theory which is disputed between Minsky, Jerry Feldman and others. However the origins of the frame problem is in Hume's Problem of Induction.

for the frame problem, neither the development of a new logic to handle it is suggested, instead I will highlight the cause of such problem, which is locality. Without assuming the locality principle, complexity is unmanageable.

The theories described in the books of physics are local in space and time, relating close spatial points and time instants i.e., the distances between points and instants are ideally infinitesimal. As I am proposing a general principle that must serve as a foundation for the modeling of cognitive systems, locality cannot be understood exclusively in terms of physical distances. Locality conveys direct links, also called arrows or morphisms. One link fixes a relation between two things, the links can be simple or composed, the former defines a direct relation and the composed link is a composition of two or more direct links. For example, in the below graph, the arrows l_1 and l_2 define local relations between two nodes.

$$\bullet \xrightarrow{l_1} \bullet \xrightarrow{l_2} \bullet$$

The composition $l_1 \circ l_2$ is depicted in the upper arrow in the graph below, which represents a global property of the graph because it encompasses the two local relations l_1 and l_2.

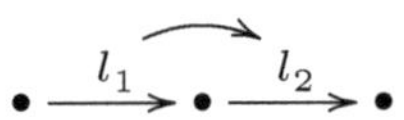

As it is pointed out in [257], evolution may be alternatively thought of and analyzed as a composition of local procedures. The same metaphor could have inspired William James when defined consciousness as a continuous stream of thought [258]. In this vein, human consciousness could be explained as the composition of local thought processes, leading to global loops of processes of processes, that result in a higher-order stream of consciousness. Dealing with local to global properties is one of the most critical problems of today's science. It is far from obvious what algebraic structures will be useful to obtain precise local-to-global results. Category theory provides the notion of colimit which can be used to deal with local to global properties. It might be noted that colimits do not necessarily exist in all categories, when it does the colimit captures the emergence of strictly increasing complexity. A colimit in a category can be seen as the binding agent in, for example, a neural architecture.

4.5.2 *Principle of Hierarchy*

The Principle of Hierarchy (PH) circumvents the realist/non-realist ontological disquisitions about the structure of the world seen above, in Sect. 4.4. The PH establishes that nature is organized in levels, and a level is a collection of things with specific properties in common. Levels are not empirically observable things existing in nature, but mathematical abstractions or sets of objects that share some properties. Following the principle of hierarchy, we can add the level l to the model shown in Sect. 4.4.1, resulting:

$$m(S)_l =< C(S)_l, E(S)_l, S(S)_l, M(S)_l >, \quad (4.5)$$

where the level l expresses a set of objects that share some properties.

Emergence is a systemic property, that is to say, it is a global property that cannot to be found in the particular components of the system. The reason to include emergent properties is twofold:

(i) Emergent properties do not exist in any of the parts of the system
(ii) Systems have different levels of description, therefore some properties or others are "visible" depending on which level we are observing the system

- l_i: is the level of things at level i (they share a set of properties)
- l_{i+1}: is the level of things at level $i + 1$ (they share a set of properties)

We say that the level l_{i+1} supersedes or is superior to the level l_i iff all the objects in l_{i+1} are composed of objects of l_i. For example, the level of the molecules of Hydrogen, l_i, is composed of objects belonging to the level of atoms of Hydrogen, l_{i-1}; then $l_i > l_{i-1}$ or l_i supersedes l_{i-1}. One example of a level that supersedes more than one level, is the level of the graphs, for example l_g, which are composed of elements from the level of edges, l_{g-1}, and form the level of nodes, l_{g-2}; then $l_g > l_g - 1 > l_{g-2}$. We can now try to define emergence formally: one object, Ω, composed of objects from levels, l_1 to L_n, is emergent when has at least one emergent property, and ρ is an emergent property iff ρ is found at the i level and not in any of the precedent levels, 1 to i – 1.

$$\neg(\Omega_{1...i-1,\rho}) \wedge \Omega_{i..n,\rho}$$

We can represent the configuration of a hierarchical system using the definition of hierarchical category as introduced by Ehresmann and Vanbremeersch in [184]. It ought to be noted that Rosen anticipated the use of colimit to model emergence in hierarchical systems "every object in the big category can be regarded as a limit of elements in the small one" [259], here "limit" should be interpreted as colimit in a proper categorical approach [260] (Fig. 4.2).

Definition 4.1. In a hierarchical category K, its objects can be partitioned into a finite sequence of levels, so that any object O in the level n is the colimit of at least one pattern P which is composed of objects that belong to level $n - 1$ or lower.

Objects in a hierarchical category can receive and send information from objects at the same level, but also at lower and higher levels, that is to say, for any two objects O_1 and O_2 in the category K, there is no particular constrain in the domain and the codomain in the set of all morphisms $Mor(O_1, O_2)$.

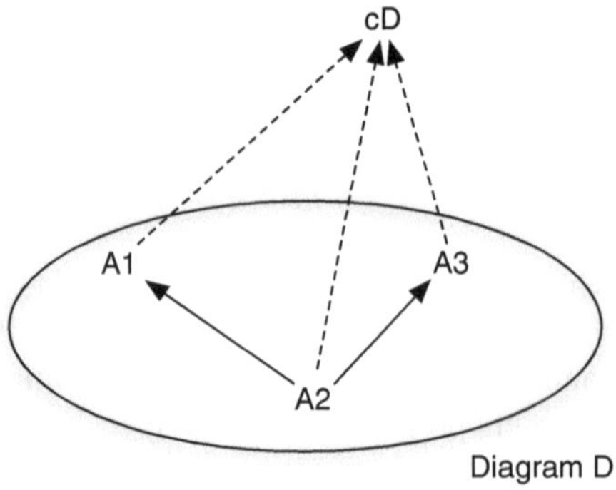

Fig. 4.2 A colimit cD for the base diagram D. For the sake of clarity, in the figure the diagram D has three objects, $A_i i = 1 \ldots 3$. The colimit shows that high level interactions i.e., the colimit cD, may arise without being directly induced from lower levels i.e.: the diagram D

4.5.3 *Principle of Multiplicity*

There is a commonplace statement that says that evolution leads to optimal systems. Such an assumption has been conveniently deflated and lay to rest in biological studies [19] and economic ones as well [261]. Evolution is not any guarantee for optimality and neither for progress. Accordingly, humans should not be delusional in thinking that we are optimal in our properties. The relationship between complexity and robustness in genetic networks, [262], bacteria [263] and other living organisms [264] is not univocal. In a theoretical study, Krakauer and Ay [265] show that robustness always implies a minimal level of network complexity. However, the mechanisms that are at the basis of robustness in biological systems are too numerous e.g., modularity, feedback, redundancy, hierarchy, distributed information processing etc. to establish a univocal causal relationship between complexity and robustness.

There exist at least three approaches – informational, computational and topological – to measure network complexity. Tononi, Sporns and Edelman [266] provide a network complexity measure based on the mutual information among all bipartitions in the network. In this view, deviations from linear distributions and homogeneous dynamics across scales, may beget an increase of complexity that can be observed in more hierarchical and modular topologies. Tononi et al. conjecture that it exists a direct relationship between complexity and degeneracy, anyhow the authors do not provide any formal theory of such hypothetical dependency. Degeneracy is defined by Edelman and Gally in [19] as "the ability of elements that are structurally different to perform the same function or yield the same output". The capacity for the biological systems to degenerate is capital in the process of natural selection. Degeneracy, if not an ubiquitous biological property, it is undoubtedly a property that exists at very different levels, from the genetic to the behavioral, passing by the cellular. And not only in biological systems, if we follow the definition given by Edelman and Gally. It is perfectly possible and indeed habitual, to find different objects with differences in their composition that carry out the same function; in other words, objects with disparate composition may have identical behavioral or

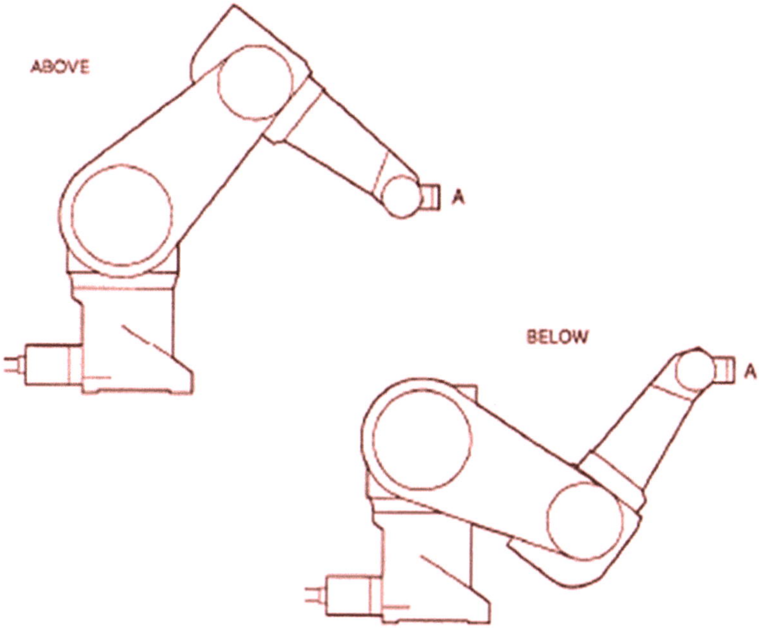

Fig. 4.3 The manipulator arm shown in the figure is an example of degenerate system if we follow the definition given by Edelman and Tononi in [19]. In the robotics literature the robot is said to be redundant

functional manifestations. It might be interesting to note that, if we use the above definition, a "simple object" such as the manipulator arm shown in Fig. 4.3 is degenerate, because for two different configurations of the robot – the elbow above and the elbow below – the end-effector reaches the same location A, as is depicted in the figure.

Ehresmann and Vanbremeersch in [184] introduce the multiplicity principle, which aims to set a mathematical frame to cope with the problem of degeneracy. We can find a historical parallelism with the formulation of the principle of inertia. The principle of inertia does not impose itself a priori, but it is verified by its consequences i.e., the motion of all material objects on the universe depend on second order differential equations. The multiplicity principle can be verified by the empirical observation i.e., biological species show adaptability and resilience. In order to use concepts such as adaptability or resilience, it is necessary to establish a way to measure them. The mathematical theory of categories can help to understand degeneracy in systems. Ehresmann and Vanbremeersch define multiplicity principle as follows:

Definition 4.2. A category **C** satisfies the multiplicity principle if it admits at least two patterns which are homologous but not connected.

For the formal definition of homologous pattern and related concepts, the reader can consult Chap. 4 of Ehresmann and Vanbremeersch book [184]. Here I will develop their ideas in a more informal way. In a category **C**, for two decompositions P and P' of the same object C, such that C is the colimit of P and P', then P and P' define a homologous pattern. It is important to note that this can occur without any direct link going from P to P' and vice versa. This shows that the colimit is a global property that cannot be seen at the level of the components, as it was already noted in Sect. 4.5.1. On the other hand, if C admits two decompositions that are not connected, C is a multifold object, and the passage from P to P' and vice versa is a complex switch. We are now equipped to explore the following statement: "The same function can be accomplish by two different structurally different components". This is a critical property in biological systems that can be redefined as follows: "the patterns that these different components form have the same colimit".

Emergence can thus, be understood as the objects and links that are de novo created through sequences of complexification processes that can be formalized as functor mappings between categories representing changes in the configuration of a system. These complexifications lead to hierarchical evolutive system, where at each step patterns at time t may admit or not a colimit. The emergence of an object cP as a colimit of a pattern P will ultimately depend upon two conditions: first, the components of the pattern must be coherently bound to the object cP and second, this binding must ensure that cP is functionally equivalent to the pattern operating collectively. Thus, an emergent object can be characterized in categorical terms as a minimal representation of a pattern P or colimit cP such that, in accordance with the definition of colimit, there must exist a collective link from P to cP and the links from cP to any object A in the pattern P are in one-to-one correspondence with the collective links from P to A, so that each collective link is bound into a well-determined link from cP to A. We conclude this chapter by recalling that the categorical concept of colimit provides a comprehensible mathematical expression for complexification process which is at the root of emergence [267].

Chapter 5
The Shift Towards Structure

5.1 Introduction

In this Chap. I flesh out the concept of structure and I explain the necessary link between structure and form. Scientists build models and theories that capture, in a more or less accurate way, the regularity of the world. This is possible because nature is structured. Modern physics studies symmetry by discovering the physical laws that remain unchanged when the system is viewed from different perspectives or undergone transformations [268]. Mathematics provides provable knowledge about the real world and this is due to the fact that mathematical structures deal better than anything else with the structure of the world. That is to say, there is a structure preserving mapping between the mathematical structure that models the world and the world itself. Mathematics aims to provide formal descriptions that capture the general structure in the most economic way. Set theory and category theory are two mathematical languages capable of describing a wealth of interrelationships between structures. Mathematical structures can be modeled themselves as a set of objects with certain distinguished relations and operations within a set or category. The main motivation of this chapter is thus, to formally frame the concept of structure, within a single unified framework.

5.2 Defining Structure

The Oxford Dictionary of English defines structure as *"the arrangement of and relations between the parts or elements of something complex"*. Structure is a word used in many different contexts and situations, for example in architecture structure is a man-made object fixed to a surface e.g., a building, in the social sciences, the term structure is used to refer human organisations such as social institutions or class structures, and in molecular biology, the structure of a cell refers to its

J. Gómez-Ramirez, *A New Foundation for Representation in Cognitive and Brain Science*, Springer Series in Cognitive and Neural Systems 7, DOI 10.1007/978-94-007-7738-5_5,

shape or geometric appearance. It goes without saying that in both natural and social sciences, structure is a core concept, theories, paradigms and whole research agendas pivot on the concept of structure.

Structuralism aspires to the systematization of knowledge and is concerned not so much with what things mean, but with the codes, structures and processes involved in the production of meaning. There is a linguistic structuralism, functional structuralism, structuralism in mathematics and even a structuralist film theory, and there is, *bien sûre*, a counter-structuralist programme e.g., post-structuralism and deconstructionism. The post-structuralism summons the structuralist project, based on the Greek ideal of rationality and in the systemic method of Descartes, as *tyranny of logos* [269]. Deconstruction is presented as the post-structuralist tool to resist this tyranny because it may show the undecidability of meaning for all texts. In this vein, it is key to understand Derrida's theory of repetition. The sign is a priori given at the beginning of the text, but one does not understand its meaning independently of aspects that are not necessarily placed in the reader's expected order. The sign's interpretation in the text is modified along the repetition of the sign appearances, therefore there is no beginning but a constellation of re-interpretations as the lecture progresses. The post-modernism approach is not exclusive to humanities, for example *ethnomathematics* [270] is one possible case of post-modernism in the "hard sciences".

We can define structure as the set of components of a system and their relationships. Based on this definition, one system has a multitude of possible structures. The structure is therefore contingent on the configuration of the system or to put it in a different way, the structure depends on which parts of the system are taken into account, and how the relationships between the different parts are calculated. But we can not accept this definition as definitive. Indeed, any partition of a system could be its structure. It is therefore necessary to make a distinction between relevant and irrelevant structures, otherwise we are hopelessly left with complete arbitrariness in what can and can not be the system's structure. The relevant structures are those that are preserved under transformation. Consequently, the structure must reflect the invariance through transformations. Thus, the idea is to go beyond the predominant view of structure as a geometrical shape (life sciences) or as a construction permanently fixed to Earth's surface (architecture), and translate the mathematical concept of structure into the arena of the natural sciences, in particular in the cognitive and brain science. In order to make this point clear, let us explore the habitual view of structure as a shape. The insufficiency of this approach is rooted in the fact that the shape of a system, when it has one, usually is a mere appearance, it lacks of explanatory power about how the shape came to be, or about how it will evolve in the future. For example, physicists do not think in terms of the shape of the proton. Although the proton overall has a spherical shape, its exact shape dramatically depends on the momentum transferred to the proton, so if a photon couples to a quark with spin parallel to the spin of the proton, then the shape is not spherical but similar to a peanut [271].

In biology, the extraordinary diversity of the different organs makes the view of organ as a purely morphological concept completely inadequate. The organs are, for the most part, determined by their forms (shapes) and also by their inner structure, which lies at the base of their outer shape. The concept of organ is a pure physiological concept, rather than a morphological one. Only fractals, like coastlines or snowflakes, have parts which are similar in shape to their whole. Perfect fractals, however, are very rare in nature. In conclusion, the form or external shape of systems can not be utilized as an universal feature for systems modeling, complex or not. Unlike energy, shape is not an universal property. Thus, strategies for system modeling that exploit the symmetry or similarity of shapes between the whole system and its parts can not be universal solutions.

Definition 5.1. Structure is the internal organisation of a system.

Proposition 5.1. *The operational aspect of structure is given by a set of objects which are invariant under transformations (homomorphism).*

Homomorphism will be formally defined in the next chapter, now it is sufficient to know that homomorphism is a structure preserving map. We will see how this definition of structure leads quite naturally to the concept of structured-system and later to the category-theoretically framed *cat-structure system*. Note that the term *cat-structure system* that will be used in this chapter is equivalent to the notion of general category that was formally defined in Chap. 3 i.e., a two-sorted system, the sorts being called objects and arrows that fulfill the axioms of associativity in composition and identity. But prior to get into the sophisticated debates around the different interpretations of mathematical structure, I will try to be as didactic as possible and I will serve myself of the bellow allegory in order to hopefully transmit to the reader the intuition of the concept of structure, which will serve for the formal treatment of the concept that will follow.

5.2.1 The Shepherd's Tale

In a remote land a shepherd lived with his flock of sheep. He never went to school, so he never learned how to count or knew about the natural numbers, let alone arithmetics. Though uneducated, the shepherd was a curious man and wanted to know if he had more or less sheep than his neighbor in the valley. At the end of the day when the sheep were back in the cave, he placed some pebbles that the shepherd collected during the day and matched each sheep with each pebble. Since then the shepherd could positively know if some sheep were missing or a new sheep had joined the group. The shepherd had arranged a mathematical structure. The sheep were now a structured system whose structure was exemplified by the amount of pebbles. This supposed a great achievement, the matching of pebbles-sheep represents the cardinality or number of members in the set of sheep. The first morphism or function between two structures had been attained.

The shepherd, as it was already said, was very curious, and after having figured out the structure of the sheep, he went further. Obviously the sheep were not all the same, our friend knew very well who was the most lazy, the one with the darkest skin, the one that most easily got distracted etc. As before, he put pebbles in the cave where he kept the sheep, but in this occasion each pebble matched each particular sheep, for example the biggest pebble matched onto the biggest sheep, the most grained pebble represented the least gregarious sheep and so on. Now the shepherd could know, not only if some sheep were missing, but also which particular sheep, simply by looking at the unmatched pebble. Therefore, the structured system is now more complex than before, when it only captured the cardinality of the set of sheep. The structure exemplified by the pebbles with different aspects, captured different physical appearances, moods ... of the sheep etc. Notice that this fable works as an anthropological explanation about how the mathematical structures started to be applied to physical systems. The same fable also have a phenomenological explanation: the cognitive agent, in our case the shepherd, describes in first person how his mind is working and finally a neurophysiologist may want to describe the shepherd's brain activity. With the tale above, I hope have made my point clear: for any physical system, the relations between its components can be encoded into a mathematical structure. For example, the structure of a pack of six can of beers is 6 and the structure of the 12,349 telephone poles that connects Berkeley with San Bernardino (CA) is 12,349. Obviously 6 and 12,349 are not the only possible structures of those systems, but undeniably they are admissible abstract forms of those physical system. That said, we can not dream that just by matching physical systems with numbers we can always get some relevant knowledge about the world. As a matter of fact, physical systems may present non empirically perceivable patterns and intricate structures that require more complex mathematical structures than numbers.

5.3 Categorizing Structured Systems

As it has been showed in the above allegory, an abstraction e.g., a number, is a mental construction that makes it possible to talk about objects without pointing to them. Abstractions function as a toy model of the world. Thus, in order to interact meaningfully with the world we impose abstract forms on it. The abstractions are our interfaces through the world [272]. The cognitive agent establishes sets of equivalence or classes between the systems (things), and the structured systems (abstractions), this is how it grasps the meaning of the world. It might be emphasized that structures can not be created from nothing, *ex nihilo nihil fit*, there must exist a pre-existing order. Michael J. Katz [273] coined the word *templet* to refer to the pre-order that makes the perception and the following abstraction of a physical system possible. *Templets* are the necessary blueprints or natural patterns for perception. They collect the configurational information that is embodied in the physical systems.

The ontogenesis of the natural patterns, or in other words, how the pattern of a system, P_S came about, are metaphysical questions. Metaphysics, as much as other

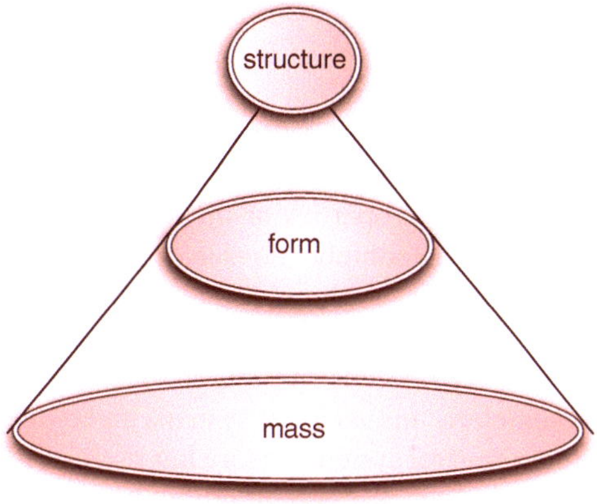

Fig. 5.1 *Structure* is a kind of *form*, while *form* is one possible kind of *mass*

forms of constructing metaphorical worlds, is of little or no help in building state representations for system's prediction and control which is the aim of scientific models [274]. We must be prudent in the use of speculative or metaphysical disquisitions about what a pattern in the physical world is. We should ask which formalism can describe more accurately the patterns and systems existing in the physical world, or put in other words, how good is the mapping that relates physical systems with mathematical abstractions (Fig. 5.1). In what it follows we try to shed some light on these two questions: given the pattern P_S of the system S, how can we obtain a model or representation of the system S, and what is a structure preserving form of S?

5.3.1 Structured Systems

We must start by mentioning the two main interpretations of structure in mathematics: "in re" and "ante re".

- "in re" mathematics is about systems with structure or *structured systems*, rather than about the structure itself. The shepherd's allegory is indeed an "in re" interpretation, the numbers are a shortcut to talk about systems (sheep) that exemplify the structure. Thus there is no structure as such but structure needs to be exemplified in a system.
- "ante re" mathematics assumes that structures exist over and above systems that exemplify them. Thus, the "ante re" interpretation reifies structures because they exist independently of any given system or object, structures are prior in place, in this view, structure is reminiscent to Platonic forms.

The mathematical structuralism of Bourbaki assumes an "ante re" conception of structure, that is, structures are entities on their own [275]. Thus, mathematical structures are constructed by appending to an initial abstract set additional constructs, where the abstract set exists prior any component is added. The systems theory approach is, in contrast with Bourbaki, "in re" structuralism. In a typical systems theory discourse, structures are subordinated or even eliminated in favour of systems. Indeed, structure is here a shorthand for talking about systems of objects [276]. The models built by systems theorists, rely on the measurability of the relevant properties of the system, namely the variables of the system. There are two main complications in this approach. Firstly, it is not clearly stated what the real nature of the properties of the system are, neither which are the most relevant properties. Secondly, the definition of system in systems theory is too vague and general. A rigorous analysis about whether the systems are ante re or in re objects, is somehow missed in the systems theory literature. Whether or not we want to admit it, the general systems theory has more of a curiosity in the history of science books than of a scientific program that progresses and makes quantifiable contributions. Sadly, the general systems theory never had a major impact in either the scientific or the philosophical context. This is not the best place to argue about the obliviousness of professional philosophers and academic scientists with regard to the general systems theory program. For a deeper review in this topic [277]. However, systems science is these days a dynamic and evolving field. While there is no so far a clear procedure or specific roadmap for the actual implementation of such a program, relatively recent fields such as systems biology aims at a system-level understanding of complex biosystems [278, 279].

In this book, the term system is used as equivalent to the term *thing*, that is, the systems are the matter of which the physical world is made. Thus, the things of the world are systems. As the world is nomologically ordered, the systems are undoubtedly structured. The term *structured system* is therefore an abstraction of a system, that is to say, models of the things that exist in the physical world. Thus the structure of a system is the abstract form of a system. Any system, when perceived by a cognitive agent, is "naturally" structured. If the system lacks structure, the problems disappears, not because we solve it, but because we erase the statement of the problem. A system without any structure at all can not be a system. There are underlying structures which are present in the physical world, these structures may be common to mathematical structures and therefore can be formulated in the mathematical language of category theory [280].

In a very abstract i.e., categorical way, a type of structure can be specified as having:

1. A class of objects (or maybe more), for example in a graph type of structure there are two classes of objects: nodes and edges.
2. A class of structural maps that conform to a specific abstract specification, for example in a graph type of structure we find two structural maps i.e., *source* and *target*

A dynamical system can accordingly be seen as a type of structure with one component object of its states and one structural map given the equation(s) that describe its dynamics. It ought to be remarked that an abstract specification of a type of structure can be considered a graph [172]. A (directed) graph (sometimes

$$N \underset{t}{\overset{s}{\rightrightarrows}} E$$

Fig. 5.2 A graph is a set of nodes N and a set of edges E, together with two functions s (source) and t (target)

called a network) is a pair of sets of nodes and edges (N, E) together with a pair of assignments *source and target* or s and t for short. The abstract notion of graph structure is illustrated in Fig. 5.2.

Note that for any graph G the nodes may be identified with the graph map $N \overset{\alpha}{\to} G$, the arrows with the graph map $E \overset{\alpha}{\to} G$, the source of α is $s(\alpha) = s \circ \alpha$ and the target is $t(\alpha) = t \circ \alpha$.

5.3.2 Structured Systems as Categories

Category theory provides, a "in re" interpretation of mathematical structuralism. Categories are not just a new fancy mathematical abstraction that means the same as set or class, rather it provides a theoretical frame for the study of several structured systems related in different ways. It is interesting to compare this categoric approach with the set-theoretic interpretation of structure which has been the predominant view, at least in the period in which Bourbaki school was the most strong voice in mathematics and very specially in the didactics of mathematics. For Bourbaki, structures are types of set-structured systems. Contrary to set theory, in categories, the objects do not need to be elements and the arrows do not need to be functions. Categories are more general than sets and than any other mathematical structure. Category theory is able to formally define and operate with both structures and structures of structures [281], that is to say, it provides a general framework for defining things or systems in our terminology and ways to having maps between things. Typical mathematical structures e.g. sets and functions, groups and homomorphisms, and topological spaces and homeomorphisms are zero dimensional. A cat-structure is a mathematical structure that can be modeled as a category, Chap. 3 provided a number of examples. More interesting is to exploit the fact that category theory is inherently hierarchical. For example, one has in ascending levels of complexity: object, arrow, category, functor, natural transformation, adjointness, monad, higher-order categories etc.

The category of all sets **Set** is one-dimensional, and following right along, the category of all categories **Cat** is "two-dimensional." and the ascent continues on to (n + 1)-dimensional categories [282]. Thus, cat-structure is iteratively defined as 0-dimensional as in a structure as a category, 1-dimensional as in the category of all sets **Set**, the category of all graphs **Graph** or the category of all topological spaces **Top**, 2-dimensional as in the category of all categories etc. The evolution from to category-structured systems, to categories of structured systems to finally categories of categories is explained bellow in three steps. A *cat-structured system* of 0-dimension is a structure, for example a group, that admits to be defined as a category. A *cat-structured system* of 1-dimension has objects as structures e.g., groups and arrows as structure preserving maps between groups. A *cat-structured system* of 2-dimension is the category of all categories. It is important to note that a 0-dimension *cat-structured system* is a discrete category i.e., the arrows are identities, while 1 and 2-dimension *cat-structured system* contain a collection of arrows, different from the identity, for any two objects.

The axioms of category theory – morphisms compose associatively with identity – provides the means to talk about structured-systems without being bothered with what the objects of the systems are made of, but rather focusing on how morphisms relate to each other. The relationships between the morphisms in a category or between the functors in the category of categories provides the patterns that allow to analyse the shared structured between systems. The axioms for a category provide the context from within we can analyze the shared structure of abstract kinds of structured systems in terms of the morphisms that exist between them [283]. The important point to be retained here is that it is possible, using this perspective, to talk about shared structure of structured systems in terms of types of cat-structured systems. The nature of objects and arrows in the category-structure system is practically irrelevant, what really counts is the patterns of morphisms in the category. Thus, a category acts as a schema that is used to frame the statements we can make about the shared structure of systems, in terms of types of cat-structured systems.

5.4 An Example of Theoretical Hypothesis in Biological Systems: The Brouwer Theorem

The sections above facilitate a review of the concept of structure and paves the way for its mathematical formulation. In what follows, we will try to apply it to real-world problems. It is worth reminding our selves, that a central concept of this book is that of structure. The idea is to provide a framework that deals with objects as relational components in structured systems by way of their shared structure, rather than as clear-cut systems that posses relevant properties to be measured. In this section, it is exemplified how pure mathematical statements may stimulate the research in areas that are apparently disconnected to the mathematical realm e.g., spatial representation in the brain.

The Brouwer's fixed point theorem was originally created for topological spaces and continuous maps. I begin by giving a formal definition of the fixed point theorem, next an original use of this theorem is provided. The rationale is to show that the Brouwer's theorem may illuminate the place-cell/place-field mapping problem. The place cells are nerve cells in the brain hippocampus that encode the position in which the animal is located at a certain time. Thus place cells are active in their respective place fields [232].

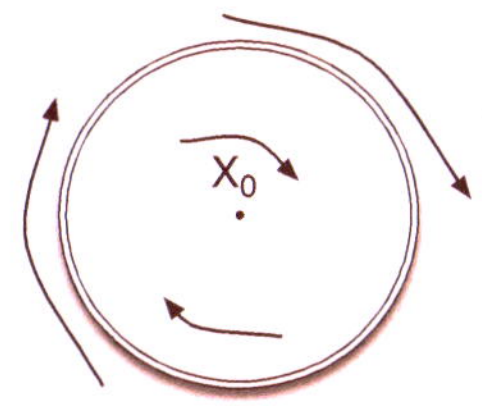

Fig. 5.3 Rotating the disk gives a continuous endomap $f : D \to D$. The *center* of the disk is the fixed point x_0

Theorem 5.1. *Let I be a line segment $I = [a, b]$ and $f : I \to I$ a continuous endomap,*[1] *then there is a fixed point x_0 in I such that $f(x_0) = x_0$.*

This is the simplest version of Brouwer theorem because the continuous function f is defined in the one dimensional space I (a line). For a more general form of the theorem we need to take the mapping f as $f : K \to K$, where f is a continuous function from a convex compact subset K of Euclidean space to itself. An example of K is the disk D in Fig. 5.3.

Thus, f maps topological spaces, that is, f is necessarily continuous, so a small change in the domain corresponds with a small change in the co-domain. Therefore no big leaps are permitted in the function f. From the definition it follows also that f does not preserve distances, that is, it does not define a metric, rather it preserves relative distances. For f to be metric-preserving is should satisfy that for every metric space $(\mathcal{X}, d)$, $f \circ d$ is a metric on $\mathcal{X}$. It is easy to see that this is not the case as the condition, *for all* x, y, $f(x + y) \leq f(x) + f(y)$, does not hold.

Let us see now how to make use of the theorem in the hippocampal place cells. Two requirements are in order here. First, as f is an endomap we need the same category at both sides of the function, and second, f needs to be continuous. Let us imagine now the experimental setting. We want to explore how the place cells in the hippocampus of a real rat represents the environment in which it has been placed. To make the experiment easier, the environment is one dimensional, that is, an animal, for example a rat, can move in only one direction in a rail track. Suppose that f represents the mapping between the place cells and the physical positions in the rail track visited by the rat, as f is an endomorphism, it takes values in the same space in which it is defined, $f(x) \in I$ for all x, where x is the location of a place cell in the hippocampus. Hence, if we find a setting in which the assumption above holds, the fixed point theorem applies, then there is a fixed point. But in the rat's brain, such a continuous function f has not been found, as a matter of fact, close pyramidal neurons in the hippocampus do not necessarily represent close points in the environment [43]. Thus, as the hypothesis of Brouwer theorem does not hold – f is not continuous – we can not say anything about the possibility to have fixed points $f(x_0) = x_0$, which is the definition of place cells using the Brouwer notation.

[1] An endomap or self-map is such that the domain and the codomain are the same.

Table 5.1 Gedankenexperiment of the Brouwer problem

	$H \to C$
if $\neg C$ then	No inference about H
if $\neg H$ then	C can be true or false
if H then	C is true

However, from the analysis of the Brouwer's theorem, we can still infer important consequences that are sketched in Table 5.1. Brouwer states that

$$\text{If } H = f : I \to I \textit{ is a continuous endomap in the interval } I \text{ Then}$$
$$C = f \textit{ must have a fixed point } f(x_0) = x_0.$$

Logically, if C is true then we can not say anything about the validity of the hypothesis, H, which can be true or false. Experimental research is focused on factual data which is a very fine praxis as long as the theoretical aspiration to put the data together in a sound theoretical body is not neglected. For example, given that we have discovered place cells, that is, C is true, we should try to incorporate this fact into a sound theoretical scheme, for example, the Brouwer theorem. On the other hand, if H is true then C is necessarily true, then provided that there is a continuous function, the fixed place cells exist. If H is not true, then we can not endorse or reject the statement "fixed place cells exist". These three cases are sketched in Table 5.1.

So far, critics would justly point out that to use Brouwer's theorem is an exotic attempt that does not reveal any deeper insight about how hippocampal place cells encode location representation. This is true, especially if the research is driven by strict experimental analysis or computational simulation modeling. In the first case, the research's aim is to collect data in order to validate or reject hypothesis, and in the second case, the target is the reconstruction by computational simulation of biological phenomena. Experiments with rats in a linear track show that different neurons are activated when the rat visits the same position, in this light the concept of place cell as used above needs to be, in the best case, reformulated or even discarded. This is because in order to apply Brouwer, both the brain metric and the environment metric need to be commensurate. On the other hand, computational reconstruction of place cells by neural networks able to learn physical positions, can only hardly provide useful knowledge far from the particular requirements for which the simulation was planned.

The idea behind the introduction of the Brouwer theorem here is simple, rather than corroborate hypothesis or simulate phenomena, the goal of this approach is to generate new hypothesis. For example, let us suppose that the hypothesis H is true for a given system e.g., a brain hippocampus or a bio inspired robot with an on board navigational system based on the hippocampus of the rat. In this case, and for direct application of Brouwer's theorem, the hippocampus of the system has one or more fixed place cells. From the biological standpoint, it seems questionable the utility of building an artificial "hippocampus" such that it exists a continuous function between the hippocampus and the space. Admitting that such a scenario is not "realistic" that is to say, is not true, however, it may suggest new research. For example, in most of hippocampal place cell models it is assumed that place cell firing is encoded during first exploration and later consolidated during rest or sleep. However, Dragoi and Tonegawa [284] have shown that the firing of place cells can also occurred during the resting or sleeping period to encode novel experiences occurring in the future. This is the "preplay hypothesis" by which the animal during sleep anticipates places that will visit afterwards in the awake state. This kind of discoveries rely upon a hypothesis driven approach, in which hypothesis apparently incomplete or counterintuitive can be empirically tested and proved correct.

Chapter 6
A General Framework for Representation

6.1 Introduction

In this chapter I present a general framework for representation based on category theory, which was previously introduced in Chap. 3. The idea is to bring a new mathematical formalism into the domain of representation of physical spaces, setting the basis for a theory of mental representation, able to relate empirical findings, uniting them into a sound theoretical corpus. The major benefit of the application of this theory based on category theory is that, on the one hand, it may help to discard conjectures that are at odds with this formal framework, and on the other hand, it will facilitate the integration of different models of representation into a durable theoretical framework.

The chapter is structured as follows. Section 6.2 come to grips with the triadic nature of representation. Section 6.3 discusses the concept of representative models and sketches a general theory of representation in cognitive systems. Specifically, it is argued that representational models must carry within them some sort of structural commonality e.g., similarity, isomorphism and homomorphism. In Sect. 6.4, a general theory of representation based on category theory is introduced. The main rationale of this theory is that representing an object is to define a collection of morphisms by which the relationships between components of the object can be precisely specified. Moreover, examples and applications of the theory are given in both the category of metric spaces and the category of neurons.

6.2 Representation is Triadic

We all know of the existence of symbols, for example, icons, pictures or words that represent real and abstract things. With the use of symbols we can talk about objects without having them in sight. But symbols do more than simply denote other objects, let us see why. Representation is ordinarily assumed to be the same

J. Gómez-Ramirez, *A New Foundation for Representation in Cognitive and Brain Science*, Springer Series in Cognitive and Neural Systems 7, DOI 10.1007/978-94-007-7738-5_6,

Fig. 6.1 The image features a physical chair, a photograph of that chair, and the text of a dictionary definition of the word "chair". Both the photograph and the text are models that allow its user(s) to infer specific properties from the model to its physical embodiment. The image corresponds to "One and Three Chairs" (1965) by the conceptual artist Joseph Kosuth's (Source: Tony Godfrey, Conceptual Art, London: 1998 [20])

as denotation. For example, the red traffic light represents stop, or the word "MAN" for an English speaker, means a human male. But denotation is not the only way to establish representations. In fact, denotation is a weak version of representation, this is because anything can denote anything. Denotation is just a matter of convention, the red traffic light in some railway networks mean "clear" rather than "stop", and the word "MAN" can denote a brand of a truck. I disagree with the philosopher Nelson Goodman who claims that "denotation is the core of representation and is independent of resemblance" [285]. Denotation is far from being the core of representation, indeed denotation is a non-epistemic representation.

A model, m, is an epistemic representation of a system, S, when the user of the model m can perform sound inferences from the model that are applicable to the target S. This is also called surrogative reasoning [286]. For example, while a map of the metro is an epistemic representation, the logo of the metro is not i.e., the user can make surrogative inferences from the map to the real network and not with the logo. In conclusion, in denotation, someone uses the model to represent the system, while in surrogative reasoning, the model allows its user to perform specific inferences from the model to the system. This has an important consequence: either denotation and epistemic representation define triadic relations and not dyadic, as it is usually assumed. A photograph or a map are representations of a target for a user or set of users. Thus, there are three elements involved in the representational relationship i.e., the vehicle (the photo of a tree), the target (the tree) and the cognizer (a situated interpreter). Accordingly, a symbol is not a representation in itself, but is always a model for some cognitive agent, in some context. The artist Joseph Kosuth has explored this idea of representation in "One and Three Chairs" Fig. 6.1.

Figure 6.2 shows the process of perception as a triadic symbol-concept-referent relationship. The external world referents are grasped by the agent's mind as concepts and can be externally expressed through symbols. This is because the world is populated by material things which undergo processes which emit energy to be captured and encoded by sensors Fig. 6.2(1). The sensory stimuli captured by the agent are objective and quantifiable. The properties of the perceived object can be measured, of course, the agent has perceptual limitations about what can and cannot be perceived, based on the nature of its sensors and the way they are attuned. The patterns are instantiations of the concept's properties for certain kinds of perceptions Fig. 6.2(2) that try to match up with the encoded information in the sensory channels Fig. 6.2(3). When this computation succeeds, the referent is incorporated into the concept ontology. In other words, the salient features or properties of the referent are identified and related to the agent's ontology of concepts. The conceptual component of a sign is depicted in Fig. 6.2(4). In fact, it is an ontology of concepts which represent things or processes with common properties. According to this, the ontology of concepts is nomologically related thanks to the relationships among the properties of the concepts. Due to the lawfulness of the concept relations, learning is possible; if the cognitive agent lacked an ontology of concepts, it would have scattered options to survive. Alternatively, if the agent as is the case in humans, has a language or some other sign-denotative system of symbols, the relation between the external referent and the ontology of concepts can be bypassed with a symbol. The symbol Fig. 6.2(5) serves as a vehicle to share concepts within a community of agents. However, there are other symbols that are not merely denotative. Instead, they permit us to infer, validate and even create novel knowledge. We call such symbols, models.

Now, how do we know what is and what is not a model? The search for the essential features that make something a model seems a futile exercise. As a matter of fact, cognitive agents produce a model by using it as such [287]. We may also consider, how good or representative a model is. The practice of science can be equated to the practice of building models of parts of the world. Scientific models, provide statements which are empirically refutable, that lead, to a certain degree of accuracy in prediction and/or phenomenic explanation. It goes without saying that the search of a perfect model of the world is essentially misguided because the perfect model of the world is the world itself.

The concept of model is too widely used to pretend to formulate a single definition that pleases everyone. However, there is a general agreement in that a model represents aspects of the world. In adopting this statement as a guide to provide a definition of model, the problem now is to understand what representation is, that is to say, in virtue of what specific feature(s) a certain model represents a certain system? [288]. But the concept of representation is, alas, as elusive as that of a model. A model m is adequate for modelling a real system S, when S can be replaced by m for some purpose. Thus, models stand for a primary object that we need to know, in terms of a secondary object, the model itself. For example, having a model of the Golden Gate bridge, permits us to calculate how many cars can be simultaneously crossing it, without causing any damage to the bridge. But how adequate can models be? It is thought that a good model is one that predicts the future states of the real system. This argument, apparently sound, fails when we realize that Ptolemy's model of planetary motion was able to predict the path

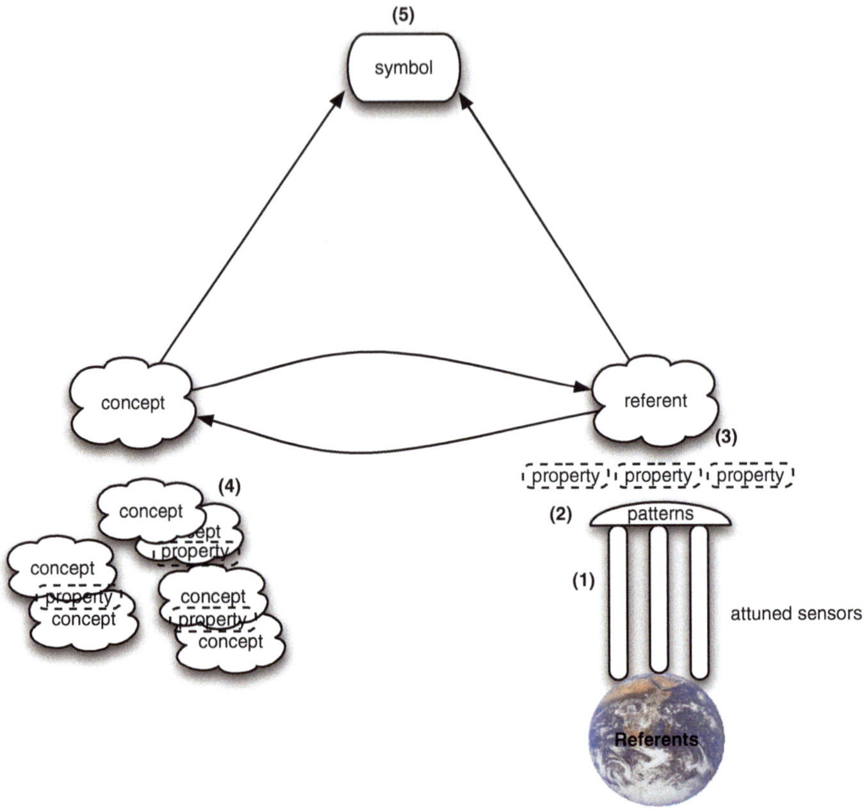

Fig. 6.2 The triadic representation conveys to us a new definition of representation as *the process of construction and relationship of signs within a system of signs*

followed by the sun and the moon. The model's adequacy is inherently linked to the purpose with which the model has been constructed. The model can be built only to show some similarity with the real system, or it can be built in order to make the model and the real system interchangeable, in this case we say that the model m is isomorphic to the system S or $m \simeq S$. Carver [289] distinguishes between phenomenally adequate models and explanatory models. The former are those that subsume the phenomenon under observation, and the later are those that constitute an explanation because they afford the control and manipulation of the real system. The ability to control a system involves the understanding of the system for different initial conditions. This is what in [290] is referred as a modeller able to answer *w-questions* of the system. A *w-questions* is a "what-if-things-have-been-different" kind of question.

6.3 A Theory of Representation in Cognitive Systems

Ronald Giere, a philosopher of science and former physicist, has worked on a unified cognitive theory of science [291, 292]. This theory is interested in a particular type of representation that is established between the scientist and the (scientific) models built by him or her. The rationale behind this approach is that given that a theory of a phenomenon helps us to explain the phenomenon itself, a theory of science has to explain the phenomenon of science itself. Thus, the task of engaging in scientific inquiry is considered as just another cognitive activity, susceptible to be studied in the same way as any other mental activity. Giere [293] succinctly defines his theory as:

AMRWP

that stands for: an **A**gent intends to use a **M**odel to **R**epresent part of the **W**orld for some **P**urpose. The expression contains at least two terms that need to be carefully considered. First, at the center of the formula, **R** or representation. Second, the inclusion in the formula of the purpose, **P**, implies an intentional conception of representation. In the previous section we argued that anything can be potentially a representation of anything else, the point we want to develop here is that only those representations that embody relations that are preserved between the represented and the representing are genuine *representational models*.

Definition 6.1. A *representational model* is a structure preserving map between the specification of the model (vehicle) and the represented object (target).

6.3.1 Structural Commonality in Representation

There are two main arguments against a naturalistic and formal account of representation. One is that representation is a relation between two different domains that need to be interpreted by an user or observer. The second states that representation is context dependent, therefore we can not get a formal definition of representation because the phenomenological experience of the agent and the context in which this agent is immersed would be missing. Thus, in order to naturalise representation, we must find a kind of relation of representation free from the agent's purposes. A relation of representation between two domains implies an agreement in form. Structure is the abstract form of a system, it allows us to focus on the relationships among the objects, and ignore other features of the objects that do not affect how they relate each other [294, 295].

Cognitive agents do not have a total or omniscient access to the real structure or the world, rather they operate within it through the use of a continuous and dynamic endorsement between the form exhibited by the system and that idealized in the agent's model [287]. To build a representation is intrinsically formal or structural, this view is opposed to the idea of representation based on linguistic definitions. The thesis defended here stands opposed to the widespread and long-standing approach to representation in terms of linguistic or pictorial symbols, instead it proposes a formal mathematical shift in representation. In order to state that A represents B there must exist a transfer of structure between the two different domains. Thus, if A and B share a structural commonality, then A represents B or B is represented by A. But what is this structural commonality? Suppose we have answered this question, in that case, we have obtained a naturalised formulation of representation, that is to say, the purpose of the agent has been naturally dissolved in the structural account of representation. Thus, in the formula proposed by Giere shown above, the term **P**urpose disappears. Now,

$$\boxed{\text{AMRW}}$$

There are three kind of relationships as possible candidates of *structural commonality*:

1. Similarity: A represents B iff A is similar to B.
2. Isomorphism: A represents B iff there is an isomorphism between A and B, that is to say A and B are two isomorphic structures.
3. Homomorphism: A represents B iff there is a homomorphism between A and B.

The rest of this section is devoted to the study of these three kind of relationships. Similarity is examined in Sect. 6.3.2, isomorphism in Sect. 6.3.3 and homomorphism in Sect. 6.3.4. At last, in Sect. 6.3.5 is analysed which is the most appropriate model of representation.

6.3.2 *Representation as a Relation of Similarity Between Structured Entities*

Given A and B, we say that they are similar iff they share a subset of properties as specified below:

1. Similarity is a reflexive relation because A is similar to itself.
2. Similarity is a symmetric relation. Let A share the properties p_1 and p_2 with B. Then B share with A p_1 and p_2.
3. Similarity is a non transitive relation. Let A share the properties p_1 and p_2 with B. Let B share p_3 with C, then A and C do not necessarily share the same property with B.

The concept of similarity presented here is more general than that of similarity as visual resemblance. Indeed, two items can be "similar" following the above definition even if they are visually disparate.

6.3.3 Representation as a Relation of Isomorphism Between Structured Entities

Given A and B, we say that they are isomorphic iff there is a one-to-one and onto map between all the components of A and B. Therefore, a map $f : A \to B$ is an isomorphism if exists the inverse of f, f^{-1}, such that $f \circ f^{-1} = 1_A$ and $f^{-1} \circ f = 1_B$, where 1_A and 1_B are the identity maps of A and B respectively. Thus, once we have fixed an isomorphism $f : A \to B$ we can treat A and B as the same object because we have available the maps or arrows f and f^{-1}. It is also possible to define a isomorphic map $f : A \to B$ it has certain properties:

1. Reflexive: A is isomorphic to A.
2. Symmetric: if A is isomorphic to B, then B is isomorphic to A.
3. Transitive: if A is isomorphic to B and B is isomorphic to C, then A is isomorphic to C.

Obviously, an isomorphism describes a deeper level of "similarity" between objects. If there is an isomorphism between two objects, there is also similarity, the contrary is not true.

In category theory many constructions are uniquely defined up to isomorphism, what this essentially means is that there is no reason to, for example, distinguish between two isomorphic objects in a category, since what really matters from a categorical standpoint is the way objects relate to other objects and isomorphic objects relate in the same (isomorphic) way.

Definition 6.2. A subcategory **D** of a category **C** is a representative category if every object of **C** is isomorphic to some object of **D**.

6.3.4 Representation as a Relation of Homomorphism Between Structured Entities

A mathematical representation of a non mathematical realm occurs when there is a homomorphism between the non mathematical system and the mathematical system. It ought to be noted that Algebra textbooks do not give a definition of homomorphism *tout court*, rather they define homomorphisms between, for example, posets i.e., order preserving maps, vector spaces i.e., linear maps., groups, Lie algebras etc. [296]. From the categorist's point of view once we have the structure of interest and the homomorphisms within we can build categories. For example, monoids and their homomorphisms define the category of monoids **Mon**, vector spaces and their homomorphisms define the category of vector spaces, **Vec** etc. This is in essence what Arbib and Manes call the naive approach to categories [173].

Homomorphisms are thus, structure preserving maps and therefore they constitute true representations. Given a type of structure and a map from A and B both with the same type of structure, $f : A \rightarrow B$, f is a structure preserving map if when we replace every element a in A we get the corresponding structure in B. For example, given a poset – a set equipped with the structure $\leq$ – we say that the map between two posets $f : P \rightarrow P'$ is a order preserving map if $p_1 \leq p_2$ implies $f(p_1) \leq f(p_2)$. Homomorphism generalizes the notion of isomorphism. An isomorphism is a homomorphism that is one-one and onto.

Definition 6.3. Homomorphism is a structure preserving map between two structures. Formally, $\Phi : (A, *) \rightarrow (B, \#)$ is a homomorphism if satisfies:

$$\Phi(a_1 * a_2) = \Phi(a_1)\#\Phi(a_2), \forall a_1, a_2 \in A$$

In category theory a homomorphism of categories is called functor i.e., a morphism of categories preserving the structure, namely identities and composition. Formally, a functor $F : \mathbf{C} \rightarrow \mathbf{D}$ consists of a map $X \rightarrow F(X)$ from the objects of $\mathbf{C}$ to those of $\mathbf{D}$ and a map $f \rightarrow F(f)$ from the arrows of $\mathbf{C}$ to those of $\mathbf{D}$ such that $F(id_X) = id_{F(X)}$ and $F(g \circ f) = F(g) \circ F(f)$. Thus, a homomorphism is just a map that preserves composite and unit. Moreover functors preserve all commuting diagrams, hence, in particular, isomorphisms.

Example 6.1. A morphism Φ from a commutative ring A into a commutative ring B, $\Phi : A \rightarrow B$, is a homomorphism iff it satisfies that

$$\Phi(a + b) = \Phi(a) + \Phi(b) \; and \; \Phi(ab) = \Phi(a)\Phi(b) \; \forall a, b \in A$$

Thus, the homomorphism of commutative rings preserves the two operations defined in that structure i.e., addition and multiplication.

6.3.5 Representation Implies Structural Similarity or Homomorphism

Suárez in [288] answers negatively to the question, can scientific representation be naturalised? The rationale is as follows: representation is non reflexive, non symmetric and non transitive, therefore the similarity relation can not be representational because as we saw in Sect. 6.3.2 is reflexive, symmetric and non transitive. The same goes for isomorphism, because contrarily to the representation relation, isomorphism is reflexive, symmetric and transitive. Therefore, as the representation relationship, in Suárez's view, lacks the properties hold in either similarity and isomorphism, the project of naturalisation of representation is futile. However, as I will justify bellow, this argument is defective, indeed the premise is false, therefore the rest does not follow.

Suárez argues that neither similarity nor isomorphism do not constitute a relation of representation. The relationship of representation to be such requires *representational force* or *essential directionality of representation* which is lacking in similarity and isomorphism. Based on this statement, Suárez enlarges the definition of similarity and isomorphism incorporating his *representational force*, redefining similarity (now similarity') and isomorphism (now isomorphism') as follows:

1. Similarity': A represents B iff A is similar to B and the representational force of A points to B.
2. Isomorphism': A represents B iff A is isomorphic to B and the representational force of A points to B.

A major allegation is in order here. The view proposed by Suárez, as long as it uses obscure terms like *representational force*, can not provide a solid basis for a theory of representation. Misrepresentation by accidental similarity is one of the favorite arguments against naturalisation of representation. Suarez provides the next example. If my friend is dressed like the pope Innocenzo X, I may think that Velázquez depicted my friend rather than the pope. Therefore, Suárez argues, this disqualifies similarity as a valid representation relation, because it leads to a wrong inference, the figure depicted is the pope and not my friend dressed like him. The problem with this argument is that it limits similarity to visual resemblance. As it was already said above in Sect. 6.3.2, similarity is not mere visual resemblance, rather it must be understood as structural similarity or homomorphism. One or more agents with analogous perceptual capabilities can trivially state that two physical objects are similar by putting them together and observing if they share a number of relevant properties. Needless to say, to compare two similar objects is banal. Much more interesting is when an agent describes a physical object or phenomenon, using another object, for example a mental object. Here, for some, there is a problem because the mental object and the physical object do not share properties, the former exists in an abstract and the last in the concrete or physical domain. This argument is flawed for one reason, it is dualist, and it dissolves if we acknowledge two simple things: first, concrete objects have properties and second, the properties are parts of the abstraction of the object. By this account, models do not apply to the world by a relation of layman's similarity, but by a more precise notion of similarity that I have coined structural similarity, and that will be mathematically defined as homomorphism.

6.4 Theory of Representation Based on Category Theory

In this section I provide a general theory of representation based on the mathematical theory of categories. The idea behind this framework is that we can represent an object in a category as a collection of morphisms or arrows such that the structure of the object is fixed in the morphisms [297]. Let us go now directly to the problem.

Definition 6.4. Given an object A in the category **A**, a representation of A (in category **A**$'$) consists of an object A' in **A**$'$ together with a morphism $A \xrightarrow{\gamma} Hom_A(A', A')$ in category **A** such that for each each element a of object A in category **A** there is a morphism γ_a from A' to itself.

It is important to note that γ is an arrow in category **A**, then we must have the object A represented as a set of morphisms from A$'$ to A$'$ i.e., the structure of the object A is given by the arrows γ_a from A$'$ to A$'$ for each a in A. Thus, the object A in the category **A**, is represented in the category **A**$'$ as a bunch of morphisms, and the representation reflects the structure within the object A, which is given by those morphisms between elements in the category **A**$'$. This definition exemplifies categorically the statement described previously: representation is a structure preserving map. Let us see this with an example.

Example 6.2. A representation of a metric space M in **Met** (category of metric spaces) in the category of neurons **Neur**, consists on an object N' in the category **Neur**, together with a rule that assigns to each element m of M a homomorphism γ_m from neural assembly N' to neural assembly N', such that the structure given by homomorphisms between metric spaces is preserved in the category of neurons **Neur**.

A theory of representation, of course, may not regard representations as static or fixed items of knowledge. In order to represent objects of increasing complexity, the theory needs to be able to manipulate and create new representations from old ones. The capacity to create new representation cases is depicted through the propositions described bellow.

Proposition 6.1. *From one representation it is possible to get two new representations by taking the product and the coproduct. Let us see first the product. Given two categories* **A**, **A**$'$*, let A be a fixed object in* **A** *with two representations of A (in category* **A**$'$*) $A \xrightarrow{\gamma} Hom_A(A', A')$ and $A \xrightarrow{\delta} Hom_A(A'', A'')$, then by Definition 6.4 for each a of A we have $A' \xrightarrow{\gamma_a} A'$ and $A'' \xrightarrow{\delta_a} A''$. Using the definition of product (Definition 3.21 in Chap. 3) we get a new arrow β_a from the product $A' \times A''$ to itself that makes the following diagram to commute.*

$$\begin{array}{ccccc} A' & \xleftarrow{p_1} & A' \times A'' & \xrightarrow{p_2} & A'' \\ \downarrow{\scriptstyle \gamma_a} & & \downarrow{\scriptstyle \beta_a} & & \downarrow{\scriptstyle \delta_a} \\ A' & \xleftarrow{p_1} & A' \times A'' & \xrightarrow{p_2} & A'' \end{array}$$

It is also possible to get a new representation of A by taking the coproduct. Thus, for the same setup of Proposition 6.1 and using the definition of coproduct (Definition 3.22), we get a new arrow α_a from the coproduct $A' + A''$ to itself that makes the following diagram to commute. Note the diagram containing the coproduct is directly created by dualizing the diagram containing the product, as shown below.

$$\begin{array}{ccccc} A' & \xrightarrow{p_1} & A' \times A'' & \xleftarrow{p_2} & A'' \\ \downarrow{\scriptstyle \gamma_a} & & \downarrow{\scriptstyle \alpha_a} & & \downarrow{\scriptstyle \delta_a} \\ A' & \xrightarrow[p_1]{} & A' \times A'' & \xleftarrow{p_2} & A'' \end{array}$$

Coming back to the example above we can extract the following heuristics:

Proposition 6.2. *For an object A in the category of metric space **Met** when it is represented on n different objects N_i' in the category of neurons **Neur**, it is possible to get two representations of A by taking the product and the coproduct, $A \to Hom_{Met}(\prod N_i')$ and $A \to Hom_{Met}(\coprod N_i')$, respectively.*

Now, let us see how to get a representation of an object on a subobject. Given the fixed object A in $\mathbf{A}$ and the representation of A, $A \xrightarrow{\gamma} Hom_A(A', A')$. Let $Q' \xrightarrow{\nu} A'$ be a subobject of A' in category **A'**, how can we obtain a representation A on Q'? By definition of representation, for every a in A, there is a morphism that takes elements of Q' to Q', $Q' \xrightarrow{\gamma_a} Q'$, then the problem here is that it may be the case that for some element a in A, γ_a can take elements of Q' out of Q', that is, the representation of A on a subobject Q' is not as straightforward as initially may seem. Thus, in order to predicate the representation of A on a subobject Q', we need first to prove that Q' is an invariant subobject.

Definition 6.5. A subobject Q' is an invariant subobject of A, when for every element a in A, there is a morphism μ_a that makes the following diagram commutes.

$$\begin{array}{ccc} Q' & \xrightarrow{\nu} & A' \\ \downarrow{\scriptstyle \mu_a} & & \downarrow{\scriptstyle \gamma_a} \\ Q' & \xrightarrow{\nu} & A' \end{array}$$

It is worth reminding that invariance is an aspect of major importance that deserves closer attention. Topology is, indeed, the mathematical discipline that studies the invariant qualities of structures under deformation and transformation. Lie algebras deal principally with the invariance in geometry and partial differential equations [298]. It might be noted that Lie algebras use the concept of cyclic permutations which is at the core of Group theory. Helmholtz was the first in pointing out that cyclic permutations of a group may explain psychological questions like perceptual invariance [299]. Exploiting properties like symmetry, it is possible to map cognitive invariance into mathematical topological ones, resulting in a class of equivalent structures in the mathematical realm. Interestingly, category theory arose in the minds of Eilenberg and Mac Lane as a general theory that deals with general forms of equivalence.

Remark 6.1. From a representation $A \xrightarrow{\gamma} Hom_A(A', A')$ and an invariant object Q', it is possible to get a new representation of A on Q':

$$A \xrightarrow{\mu} Hom_A(Q', Q').$$

Proposition 6.3. *Let A be an object in category* **A**, *A' an object in category* **A'**, *and $A \xrightarrow{\gamma} Hom_A(A', A')$ a representation of A. Now, given another object B in category* **A** *together with the arrow $B \xrightarrow{\nu} A$, then the composition of arrows $B \xrightarrow{\gamma \circ \nu} Hom_A(A', A')$ is also a representation of B*

Let us now explore Proposition 6.3 with an example using the category of metric spaces, **Met**, and the category of neurons **Neur**.

Example 6.3. In the category of metric spaces **Met**, given an object A composed of elements forming an array of equilateral triangles, an object B in **Met** together with the arrow $B \xrightarrow{\delta} A$, then B is represented as the collection of morphisms from the neural assembly N' to N', $Hom_{Met}(N', N')$, where N' is an object in the category of neurons **Neur**. The structure of B is, thus, reflected within $Hom_{Met}(N', N')$. Now, there is a mapping between two metric spaces, $f : (A, d) \to (B, e)$ if f preserves distances i.e., for all $a1, a2 \in A$, $e(fa1fa2) = d(a1, a2)$. Thus, if we define the same distance for both metric spaces (A, d) and (B, e) , $d = e$, it is evident that for any subobject B of the metric space A, $f : (B, e) \to (A, d)$, therefore any object of the grid space defined in A would be a representation of the neural assembly N'. This is empirically false because not any subobject B of A can be represented using the same set of morphisms between the neural assemblies, $Hom_{Met}(N', N')$ that represent the object A. Stated differently, a mental object and one of its subobjects, do not arise from identical neural assemblies in the brain. Therefore, from the two premises: (1) $A \xrightarrow{\gamma} Hom_{Met}(N', N')$ is a representation of a mental object A on the neural assembly N' and (2) $B \xrightarrow{\delta} A$ is a morphism from mental object B to A, it does not follow that $B \xrightarrow{\gamma \circ \delta} Hom_{Met}(N', N')$ i.e., B is represented by the neural assembly N'.

Chapter 7
Towards a Theory of Brain Structure and Function

7.1 Introduction

The eminent mathematician Henri Poincaré wrote *"The aim of science is not things in themselves, as the dogmatists in their simplicity imagine, but the relations between things; outside those relations there is no reality knowable"* [300]. The study of the relation between the external world and the central nervous system, and in particular the brain, has been approached from numerous disciplines and subdisciplines that are recognizable under the portmanteau word *cognitive*. In the past, psychologists studied brain functioning, yet did not include an examination of its internal structure. The behaviorist school, that was preponderant in the departments of psychology during a good part of the last century, accorded to the biological organism a predictable behavior in terms of input and output. In the behaviorist's view, the organism reacted with mechanical precision to the stimuli and the internal workings of the organism were considered irrelevant because it was already prefixed by the stimulus/response pair. Paradoxically, some time before Watson and Skinner gathered their behaviorist theories that disregarded the biological issues involved in behavior, the biological understanding of the brain had gained an extraordinary impulse. Ramón y Cajal using Golgi's stain technique provided definitive evidence that neurons are discrete cells.

Cajal is the champion of the "neuron doctrine" which was established based on anatomical studies done by Cajal himself using Golgi's techniques for staining neurons [301]. The neuron doctrine states that the neuron is the fundamental structural and functional element of the brain i.e., the neuron is both the basic building block and the elementary signaling unit of the brain equipped with direction of processing. The McCulloch and Pitts [73] notion of the neuron as a basic information processing unit in the brain can be seen as an updating of the neuron doctrine.

Although there still remains some technical limitations in the measurement of the brain activity, those technical barriers will eventually efface, as the resolution in the brain imaging devices is progressively enhanced and new and more

J. Gómez-Ramirez, *A New Foundation for Representation in Cognitive and Brain Science*, Springer Series in Cognitive and Neural Systems 7, DOI 10.1007/978-94-007-7738-5_7,

powerful techniques deployed. Today, the neurobiology of behavior is a reality that is continuously providing a deeper insight into the brain functioning and the neuronal correlation of behavior and cognition. Indeed, neurobiology is vindicating Hippocrates' hypothesis that the proper study of the mind starts with the study of the brain [231]. However, the models for translating the data from the micro level to the macro level and vice versa, are still uncertain. The organisation principles of the brain are still missing. The formation and disbanding of meaningful brain patterns lacks an unified theory.

This chapter explores three main issues. The first concern is an attempt to point out the difficulties – from an historical perspective – in delineating the relationship between brain structure and function. More specifically, I draw attention to the restricted, yet inconclusive quarrel between the different schools of brain organisation and functioning. Modular, distributed or sparse conceptions of brain connectivity may be verifiable or not, depending on the measurement technique and the scale that is being used. The second concern is to elaborate the idea that brain mappings have both form and meaning, suggesting practical implications of this idea to understand the connection patterns that underlie cognition. The last part of the chapter is focused on the introduction of mathematical descriptions, in particular network theory, to analyze brain networks. Some of the most important network measures, together with the challenges that the network-based approach is faced are outlined. A categorical framework that aims to shed light on the complex relationship between structure and function in brain networks is proposed. We take advantage of the capacity of categories to generalize graphs, to provide new network metrics useful for the systematic comparison between different connection networks.

7.2 Brain Mappings Have Form and Meaning

Cellular connectionism states that individual neurons are the signaling units of the brain arranged in functional groups. Although this is the established view in modern brain physiology, it still poses one important issue: What is a functional group and how they are created, transformed and disbanded?

In basic theory of sets, a relation between two sets is modelled through a function f that express a relationship between two sets, the domain and the codomain. Thus, a function $f : A \to B$, relates each element a in A to another element b in B. Likewise, when a neuroscientist investigates the neural substrate of a cognitive function, he or she is indeed carrying out the function $f : C \to B$, that relates the domain C of cognitive functions and the codomain B of the neuronal organization in the brain. In order to build the function f that relates cognition with brain areas, $f : C \to B$, we are confronted with the problem of granularity i.e., What are the unit elements in the brain, B, to which the cognitive functions, C, are mapped onto? We must note that if the neurons are those basic units, then the mapping becomes intractable and meaningless. By assuming that the space B is discrete the form associated with a related cognitive faculty is vague. In other words, the anatomical and physiological borders of the area under study for example, the right hemisphere or the frontal lobe, are too imprecise and general. Nevertheless, if *B* is thought to be continuous, then there are no modules to which we may ascribe any particular function, but instead, a daunting circuitry of neurons and synapses. In this embarrassment of riches, the *form* – the anatomy and physiology that mediates in cognition – can be effectively investigated, but the *meaning* – cognitive and emotional aspects of percepts – is missing.

Traditionally, percepts and their representations in the brain have been taken to be "covariates", meaning that the variates are essentially in conformity, being possible to establish an isomorphism between the external object and the world of the mind. Roger Shepard challenged this view. The perceptual isomorphism should not be considered as a first order relation between percept and its corresponding internal representation, but in a second order correspondence between relations among external alternatives and their projections among the corresponding internal representations [302]. Although Shepard's Second Order Isomorphism has a more appropriate psychological basis than Locke's isomorphism [303], an understanding on how our brain represents the world needs to address the neural circuitry that underlies cognition and emotion.

The nature of brain mapping must encompass form and meaning, and the best way to capture such a relationship is to focus on how the neural activity of the percept integrates both in space and time into the percept at a cognitive (meaningful) scale. It is possible, in the appropriate laboratory conditions, to trace the course from the stimulus to the neural pattern of activation triggered by it. I will explain this with two examples of increasing complexity -sensorimotor topographic patterns and meaningful patterns in the olfactory bulb. A pattern is a meaningful structure made up of an ensemble of neurons which are functionally connected. I will outline through two examples the implications of this definition of pattern.

7.2.1 Sensorimotor Topographic Patterns

Children in the school learn that we have five senses: taste, smell, sight, sound and touch. However, such classification though didactically useful, is in the words of Patricia Churchland "notoriously inept [304]". Indeed, the nervous system has receptors not only for those five senses but also for a number of other things. For example, detecting changes in the position of the head, detecting changes in the blood pressure or in the oxygen level in the arteries etc. Furthermore this classification is careless with the senses of other animals. For instance, bees can detect ultraviolet lights and a magnetic compass for positioning has been demonstrated in 18 species of migrating birds [305]. The senses are located in particular brain regions where specific type information is processed. The sensory neurons in those areas typically form specific maps of the receptor sheet of cells in the proper organ – maps of the body surface (touch), the retina (sight), the olfactory bulb (smell) or the basilar membrane of the cochlea (hear). The understanding of these maps is the first stage in the process of representation of the external world. The neural maps for sensory perception seen above, and also those representing the limbs' musculature and specific movements are arguably well-known. What is still not fully understood is the ways in which these maps are interconnected.

Features of the stimulus, for example the intensity of light or the frequency of sound, are captured by sensory neurons attuned with those events in the form of receptor potential. When the receptor potential is bigger than the threshold for that cell, an action potential is transmitted along the axon. The greater the overpass of the threshold is, the greater is the frequency of the action potentials in the axon. Similarly, the longer the stimulus duration, greater is the number of action potentials that are transmitted. The neurotransmitters released by the sensory neurons interact with their counterparts in the motor neurons and initiate a graded synaptic potential. If the membrane potential in the motor neuron is greater than its threshold, an action potential will be created which will again, cause the release of neurotransmitters that produce the synaptic potential, able to create the final action potential in the limb's muscle that results in the muscle contraction. Intentionally, the above description is a simplification, for more detailed descriptions the reader may consult any good manual of neurophysiology [231, 306]. So, apart from technical details, there is one important thing to be said. The transformations of the stimulus into neural signals that convey information are in fact, one-to-one maps, also called isomorphisms. The pattern is hardwired in the path between sensory and motor neurons. As a result of the isomorphism, the neural circuitry or pattern that will be triggered by one concrete stimulus like a stretch reflex, is quite predictable, likewise, the same stimulus will produce the same neural pattern. It should be noted that the election of the above example, the signaling network between sensory and motor neurons, is not casual. The information conveyed by the sensory neuron acts directly on the motor neuron, producing the muscle contraction. The study of the neural circuitry of sensory neurons connected directly to the motor neurons other than those in the skin, via for example visual or auditory neurons is much more difficult. When we know which are the preferred stimulus features for single neurons, we can infer the function of a local assembly of neurons from its constituent neurons, that is to say, only when we have a one to one mapping stimulus/sensory-motor neuron, the anatomical structure of the neural set will be enough to determine its function. However, there is a serious obstacle here. We live in complex environment with myriad of different stimuli to be processed by the sensory neurons. The brain map at the simple sensorimotor interaction described above is topographic, but the further we move towards associational areas of the brain in the prefrontal cortex, the less evidence we find for the existence of topographic maps. The perceptual brain is indeed topographic, but only up to the primary visual cortex or V1. Thus, only at the very first stages we find topographic maps in cortical processing of visual information (Fig. 7.1).

Cognitive operations are represented in functional maps which are implemented in neural structures that evolve over time. The challenge is to identify those maps of brain activity that mediate in cognition. On this basis, the term meaningful pattern is defined as neuronal assemblies with cognitive representational power. Stated

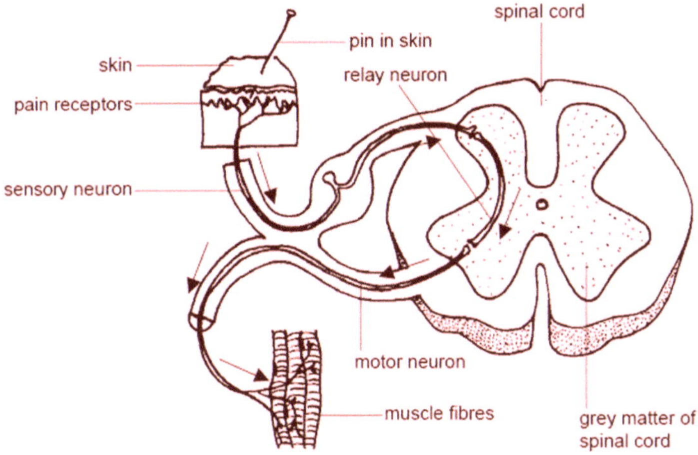

Fig. 7.1 Connection specificity – a neuron is connected with some particular neurons – and neural polarisation – within a neuron the signal is unidirectional – was discovered by Cajal (After [21])

another way, meaningful patterns are dynamic structures i.e., neuronal networks, that express cognitive operations. In the brain, these patterns are meaningful neural networks that transcend the anatomically observed modules of the brain. They are meaningful in the sense that they are functionally relevant.

7.2.2 Meaningful Patterns in the Olfactory Bulb

Walter J. Freeman, a Berkeley neuroscientist son of the famous lobotomist and remarkably gifted for electrical engineering and mathematics, has investigated the general principles of brain organisation during the last 40 years. Freeman's studies have focused on the olfactory system of cats and rabbits. Simplifying, this is how odour is sensed and processed in the brain. The stimulus – an odor – is perceived by the olfactory sensory neurons in the nose – the receptor layer. The olfactory sensory neurons project axons to the olfactory bulb of the brain. There, molecular binding to specific receptors initiates a transduction cascade that release transmitters and depolarise neurons.

The bulbar code for olfaction is spatial, this means that the odorant activity is represented in the olfactory bulb by the activation of odorant receptors with different spatial locations. With each inhalation, there are fundamentally two sorts of spatial patterns of neural activity that coexist in the olfactory bulb: microscopic which are evoked by action potentials in single cells, and mesoscopic which are informed by ensemble averaging. The microscopic pattern of single neurons, as it occurs in the sensorimotor map, is topographic and one to one, in other words, there is an isomorphism between the stimulus and neurons. On the other hand, the mesoscopic

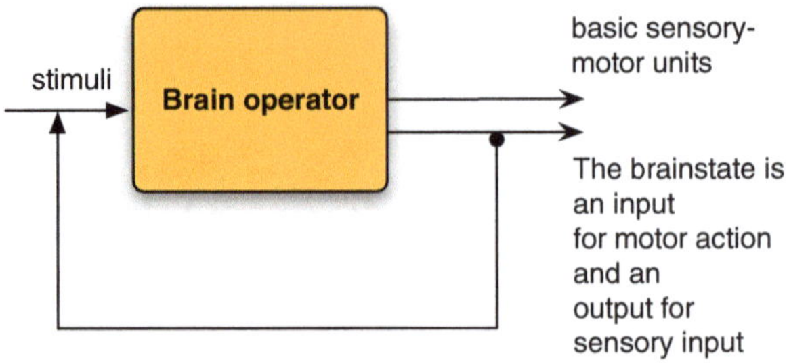

Fig. 7.2 Linear causality can tell us how a reflex response is caused by a conditioned stimulus – basic sensorimotoric loop. Thus, a stimulus excites a sensory neuron, that excites the sensory cortex, that transmits information to the motor neuron, that finally produces the contraction of the muscle. The mechanisms for perception are not strictly linear but circular. Sensory input entails motoric consequences and likely, motor output affects sensory input. In this light, the brain can be seen as an operator between the world and the body

pattern reflects how local masses of neurons organise their activity when they are destabilized by a sensory input i.e., an odorant. Thus, the bulbar patterns do not relate directly to the stimulus because each bulbar axon branches and distributes its output over the cortex. While both the micro and the mesoscopic pattern express properties of the sensed stimulus, only the mesoscopic patterns capture the meaning of the stimulus. But, how can we test this last statement?

To understand how brain activity patterns relate to the meaning of the stimulus, Freeman elaborated the following experiment [93]. A rat inhales an odorant, let us call it O_1, and after that the rat is rewarded. Then, the rat inhales a second odorant, O_2, and no reward is given. The EEG patterns associated to O_1 and O_2 are recorded. Later on, the rat smells the odorant O_1 with no reward and then the odorant O_2 is presented followed by a reward. The electroencephalography of both odorants O_1 and O_2 are recorded. The electrical pattern associated with the odorant O_1 when is presented prior to a reward and when is given with no reward at all are different, though the odorant is the same. Likewise for the odorant O_2. This shows that, as expected, different odorants give rise to different patterns, but given the same odorant again, the original pattern does not reappear, rather a new one is created. The conclusion is that the bulbar patterns do not relate directly to the stimulus but to the meaning of the stimulus. The most important consequence is that patterns are not directly shaped by the stimuli, rather the formation of the patterns is endogenous. In other words, the previous experience of the animal in dealing with previous stimuli, together with its own goals and internal motivations, mediate in the neural processing of a given stimulus. To sum up, an odorant stimulus and the elicited neural pattern in the olfactory bulb are not isomorphic, rather the measured activity patterns express the meaning of the stimulus for the subject (Fig. 7.2).

7.3 The Quest for Functional Brain Organization: Historical Account

As early as the beginning of the nineteenth century, Francis Gall developed what he thought was the mapping of the cognitive and emotional capacities into the skull and the brain areas underneath. Pierre Flourens, tested Gall's theory by trying to isolate different brain regions in order to establish the different roles played by each different brain region [307]. This was in essence the phrenologist's credo. Flourens, on the contrary, reached to the conclusion that no specific region was exclusively responsible for any particular mental operation. Rather, the brain as a whole participated in all mental operations. This holistic view of the brain was later called the aggregate-field view – specific mental functions are not localized and any part of cerebral cortex is potentially capable of performing all functions. Later on, the aggregate-field view was refuted by J.H. Jackson who, through his studies in epilepsy, showed that it is possible to locate some motor and sensory functions in particular areas of the cerebral cortex [308]. However, the aggregate-field view continued to be championed by psychologists like Ivan Pavlov and Karl Lashley. By 1950, Lashley had reformulated the aggregate-field view, based on two principles: the *Principle of Mass Action* and the *Principle of Equipotentiality* [309]. The former stated that the cerebral cortex acts as a whole in many types of learning, and the later anticipated brain's plasticity – brain's ability to change as a result of experience and recover after injury. Thus, according to Lashley's new aggregate-field theory, there was not any particular singular locus to look for in the cortex that served cognitive functions. The cytoarchitectonic view of functional parcellation in the cortex was under attacked [310]. However, facts are stubborn things and the empirical evidence for the anatomic localisation of basic brain functions accumulated. Karl Wernicke was the first proponent of the now prevailing notion of the brain as a distributed system processing. In this view, simple cognitive functions e.g., basic sensory and motor activities, are localized in particular areas of the cortex, and complex functions can be traced as the result of the interactions between simple functional sites [311]. In the 1930s, experiments with cats showed that the application of a tactile stimulus in different parts of cat's body, triggered an electrical activity in particular regions of the cortex. The mapping of the cortex as in the depiction of anatomical regions described by Dykes and Ruest [312] is a systemic account of Wernicke's ideas. In 1957 V. Mountcastle discovered the columnar organisation of the cortex, put in his own words *there is an elementary unit of organisation in the somatic cortex made up of a vertical group of cells.* However, the concept of cortical columns was not completely new, for example Berger in 1929, using the rudimentary electrophysiological techniques of the time, showed that cells of similar function e.g., primary-visual, motor and somatic, are grouped into vertical bands. For a more complete historical description see [313], and also [314–316].

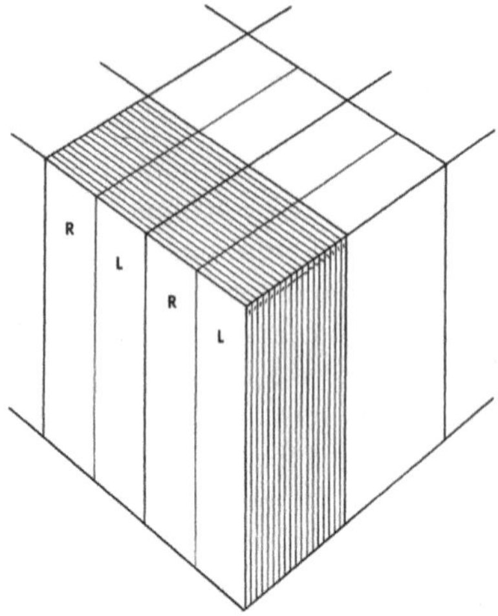

Fig. 7.3 Huber and Wiesel study of the visual field claimed that the orientation columns are 50 μm width and represent a shift in angle of 10°. A complete set of columns subserving the 180° is called hypercolumn. There are two types of ocular dominance columns – *left* (*L*) eye column and *right* (*R*) eye column about 0.5 mm wide. For purposes of illustration, Hubel and Wiesel depicted these two sets of columns as orthogonal, but there is not evidence of such relationship between them (After [22])

The idea that the cortex was arranged in the form of regular columns proved very appealing and Huber and Wiesel reported the existence in the visual cortex of orientation columns (Fig. 7.3). For both Mountcastle and Huber the cortical columns are discrete slabs with a sharp border between them. Inside each column, all the cells share some common salient property. Such a point of view has been proved to be inexact. In actual fact, columns lack borders. This can be easily noticed. For instance, the 22 orientation shifts found by Huber and Wiesel are arbitrary, the orientation mapping is continuous rather than discrete as it smoothly varies across most of the cortex [317]. While cortical columns were initially suppose to represent basic modular units of the cortex, Rakic in 1998 found "ontogenic columns" [318]. The radial unit hypothesis of cortical formation is based on this discovery [319]. As Mountcastle's original columns were larger than Rakic ones, Mountcastle coined a new term, minicolumn which was baptized as the new basic modular unit, dispensing the cortical column of that honor.

A general summary of the research in columnar organisation of the cortex is that minicolumn is here seen as the basic modular unit and the neurons within it encode similar features. The superposition of a group of minicolumns form a columnar module whose neurons share common functional properties. This might raise some concerns. First, how are minicolumns spatially located within a column? and second, is the minicolumn a functional unit of the cortex, or it is instead the cortical column who play that role? There is no common agreement on the first point. According to [320] "perhaps 50–80 minicolumns are aggregated into a macrocolumn".[1] The answer to the second point is no. The functional role assigned

[1]The term cortical column have many synonymous terms. For example, cortical module, cortical column, hypercolumn and macrocolumn are synonymous to cortical column, meaning a group of neurons which have nearly identical receptive fields.

to columns is not justified. The rationale of this is twofold. First, single neurons have multidimensional receptive fields, therefore different properties that neatly map onto a given cortical surface can never be simply detached into separate "modules". Secondly, it has been impossible to find the canonical circuit isomorphic with the cortical column. This does not disprove the existence of canonical or local circuits in the brain, rather it draws attention to the fact that it can not be as simple as clipping out a cylinder of cortical tissue in the hope that it contains such circuit. The search for meaningful patterns can not selely rely on selecting anatomical structures that are candidates to be modular units. It is rather through the understanding of the organisational principles and their formulation in mathematical terms how the connectivity patterns and their dynamics can be satisfactorily be apprehended. The point that needs to be stressed is concisely stated by Fuster *constrain a cognitive function within the scope of the method used to test it can not be a far reaching methodology* [321].

7.3.1 The Old Paradigm: Modularism-Homuncularism

Historically, attention in the identification of the structure-function map in the brain has been put on two directions, depending on the scientific background of the researcher. Neurobiologists' approach has typically focused on the identification of anatomically differentiable areas, while psychologists, on the other hand, have a penchant for sensorimotor or more complex cognitive faculties. Both approaches belong to what I call the old paradigm. But in neither of these two parallel approaches – from the neuron up to the psyche and from the cognitive function down to the neuron, the mesoscale, that is, the gap between physiology and behavior and cognitive – has not adequately been addressed.

In the human brain there are approximately 10^{11} neurons and 10^{14} synapses, these are obviously very big numbers. This daunting dimensionality problem can be drastically reduced if it is adopted a phrenologist or strict modularist approach to brain function. In this view, both the domain and the codomain in the mapping from cognitive functions to brain organisation, $f : C \rightarrow B$, are both discrete. Although such theoretical positioning can be seen as justifiable for practical reasons – it makes the problem manageable – it is empirically false. There is both a conceptual and a methodological problem in finding out which brain area a is correlated with which function f_a. When one states that area a is responsible for the function f_a, what is really meant is that the activity measured (electrical, magnetic ...) in area a has exceeded the threshold that the experimenter is using for his experiment, while other areas b, c or h have not, that is, the null hypothesis that activity in area a, $E(a)$, is less than a certain value k, $H_0 : E(a) < k$ must be rejected. However, it is self evident that having an event a occurring after an event f_a, is not the same as f_a entails a, $f_a \rightarrow a$, and of course different from a follows f_a, $a \rightarrow f_a$. However, for educational and historical reasons, in psychology, the cognitive functions have been classified in main cognitive faculties e.g., perception, memory or attention, and therefore treated as discrete items to be found in brain systems. In actual fact, those functions are not islands, but interdependent and highly coupled processes. For example, perception is function of memory and attention. If we admit that cognitive functions can overlap, their neural basis may overlap as well. But still, there is evidence that in the brain distinct functional regions exist. For example, the left hemisphere is primarily concerned with sensory and motor processes coming from the right side of the body, or the frontal lobe is related with tasks that involve planning.

Despite the evolutionary process of continuous complexification of the structure and its progressive specialisation in the brain, the different subsystems e.g., cortical areas, limbic structures etc., never act independently as isolated systems. Indeed, almost any cortical neuron or neuronal assembly can be part of many networks and thus, instantiates many different cognitive functions. Accordingly, a modularist or homuncularism [322] analysis for brain functioning must be declined as a valid theoretical framework.

7.4 Brain Cognitive Networks

Biological systems are defined by the interactions between their components [279]. It is relevant, for instance, that the biological components have an expiration date. Thus, the components of a cell at one instant t, may represent only a small fraction of the cell later on at some instant $t+k$. In order to determine the functional state of a system, we need to identify and properly described all the links in the network of interactions. Here network must be understood as a geometric object that represents the system's components as a set of nodes connected through edges denoting the relationships between those nodes. Brains are hierarchical systems, whence brain networks must capture interactions within and between brain systems. Networks are typically depicted as directed graphs and capture the result of evolutionary pressure upon the biological system. It is interesting to note that Darwin himself used this kind of formalism [164].

Inside a cell, the links between molecular components are chemical reactions or associations between chemical components. In the neural modeling, the links are synapses. Despite the gloomy complexity of biological networks – non linear interactions among highly coupled components subjected to continuous reconfigurations and structural changes driven by external forces – we can formulate general principles of connectivity. Thus, the details of the components may not be needed to understand the functioning of the system and its evolution. According to Palsson [279], the interactions are the dominant variables in the state of biological systems. However, even if we could completely characterise a biological network of interest as for example in the metabolome i.e., the full set of metabolites present in a biological system [323], the phenotype of the biological system would still remain uncertain. This is because the possible functional states in the network grow much faster than the number of its components or nodes. For example, in the human brain, a network would contain 10^{11} nodes or neurons and 10^{14} links or synapses. It goes without saying that apart from the illegibility of such diagram, practical issues like knowing which neuron is connected with which, or what is the shortest path between two distant neurons, pose problems of non computability. In order to be operational, the numbers of degrees of freedom need to be drastically reduced. Let me emphasise this point, in order to understand brain structure and function, we need to search for its patterns. This search will be futile if we do not constrain the search. What it is necessary is to surmise general principles of the system, for example: stability, robustness, adaptability. The system phenotype is embedded in those principles.

So far, I have aimed to articulate the idea that in essence, a biological system is a network of interrelated elements and that the relationships between the components play a major role. However, we must also remember that the components themselves may not be neglected. In the neuroanatomy of the cortex, the problem of how the cortical circuits work is still ontological i.e., characterize the different types of neurons, their morphologies etc. The current technologies for the morphological exploration of the brain has led to an embarrassment of riches. The few types of neurons identified by Cajal and his pupil Lorente de Nó, provided a simplified taxonomy of neurons. Today, as parameters that can be measured with new technologies increase in number and precision, the taxonomy becomes unmanageable. Attending to the different neural properties that can be measured, it is possible to characterise many different classes of neurons. For example, the worm Caenorhabditis Elegans a popular model organism counts only 302 neurons but despite this short number, White et al. [324] have identified 118 different classes of neurons in the nematode's brain. The largest class has 13 neurons with alike morphology and connectivity patterns, but many classes contain one single neuron.

7.4.1 Defining Brain Connectivity

The criterion behind our emphasis in the interactions is that the gulf between structure and function can be bridged by exploring the interplay between connectivity and dynamics. One problem with delineating connectivity is that in forming synapses, cortical neurons are promiscuous. Given that a neuron is polineurally innervated what we need most is to identify the presynaptic neuron of any particular synapse. Smith et al. [325] showed that for some layers is possible to identify a source to every synapse. This leads us at once to the question: How spatiotemporal distribution of neural activity is shaped by the underlying network connectivity?

Brain connectivity is an elusive concept that may refer to three different kinds of connectivity between brain systems: *anatomical connectivity* or anatomical links – *functional connectivity* or statistical dependencies – and *effective connectivity* or causal interactions [326]. More precisely, structural or anatomical connectivity includes synaptic connections between neurons and branching patterns of axonal arborizations connecting cell nuclei or brain regions, at the macro scale. Anatomical connectivity is a structural linkage such as synapses or diffusion channels between two neurons or groups of them. The anatomical connectivity is spatially structured. The combined activation, through synapses of discrete and local assemblies of neurons give rise to spatial connections. One anatomical local area connection may admit several functional implementations. This means that connectivity is not solely spatial but temporal as well. The arrangement of neural networks i.e. the synapses, vary during the realisation of cognitive operations. Thus, there is not a clear cut border between structural and functional connectivity maps.

Functional connectivity refers to the temporal coherence among the activity of neurons measured by correlating train spikes, or in the case of fMRI studies to temporal correlation of regional time series in the BOLD signal. To exist, functional connectivity, needs two prerequisites: one is anatomical connectivity must be already in place, and two is the existence of a function that delineates co-active states between the neurons. The more complex the functions are, for example cognitive functions, the more predominant the functional connectivity is over the anatomical connectivity. One habitual approach to understand the neural substrate that mediates in cognition, consists of identifying specific local networks that participate in cognitive operations. The function of a local area network depends on its pattern of synchronous interconnectivity with other local networks. Functional imaging techniques have led to important advances to this respect [327–330]. However, due to the poor temporal resolution of this type of technology, it is easier to establish associations between large scale areas and highly integrated cognitive functions, than to identify the local brain areas specialised in medium-low level cognitive functions.

Effective connectivity involves a causal connection between brain systems. Causality in complex systems such as the brain poses important challenges. It is thus, preferable to define effective connectivity in simple brain systems. Thus, effective connectivity can be understood as the simplest neuron-like circuit that produces some temporal relationship between two neurons. One possible example of effective connectivity is the relationship between the axon and the muscle fibers enervated by the axon, also called motor unit. Each muscle fiber is connected by one single axon, but one single motor axon innervates several muscle fibers. For example, in the human leg we find circa 1,000 muscle fibers, while in the eye an axon is connected to fewer than 100 muscle fibers. However, the effective connectivity can be traced by studying the sequence of signals that produces a reflex action and is transmitted all the way down the axon to the muscle fibers. Let us see how. When a muscle is stretched, the amplitude and duration of the stimulus are reflected in the amplitude and duration of the receptor potential in the sensory neurons. This graded signal is transformed into an all-or-none signal. But this occurs only in case the receptor potential exceeds the threshold for action potentials in that cell. The more the receptor potential exceeds the threshold, the greater the depolarization and consequently the greater the frequency of action potentials in the axon, likewise, the duration of the input signal determines the number of action potentials. The frequency and number of action potentials, is the actual information conveyed along the axon's length to its terminals and influence the amount of neurotransmitters released from the sensory neurons to the motor neurons. Neurotransmitters interact with receptors on the motor neuron to initiate a graded synaptic potential. If the membrane potential of the motor neuron reaches its threshold, an action potential will be generated and propagate to the motor cell's presynaptic terminals where the action potential causes the release of neurotransmitters, which triggers a synaptic potential in the muscle to ultimately elicit a train of action potentials in the leg muscle, which leads to the final muscle contraction. This sequence of transformation constitute an example of effective connectivity with one sensory neuron and one motor neuron.

The relationship between anatomical, functional and effective connectivity in the brain is an ongoing research agenda of the most strategic importance for policy makers in Europe and the US. The Human Brain Project and the Brain Activity Map project, respectively, are two decade-long scientific efforts that aim to completely characterize individual large-scale connectivity maps of the human brain. The interplay of the two complementary principles at work in the brain – segregation and integration – has been suggested to play an important role in the understanding of the different modes of brain connectivity described above [331,332]. Inherent in the challenges associated with the identification of functional and effective connectivity is the problem causality which is analyzed in detail in the following section.

7.4.2 Brain Causal Maps

Within the micro level of single neurons, the relevant quantities to be measured typically are the firing potential and membrane potential of neurons. Within this level there is a direct Input–Output recording that conveys a linear causality between two observed events. In other words, there is a one-to-one map between the stimulus intensity and the firing rate of sensory neurons. More precisely, there is an isomorphism between the firing rate of motor neurons and the degree of contraction of the muscles. Models at neuronal level benefit from this linear causality which in part explains the existence of topographic maps between cortical and sensorimotor components. The micro level approach, is based on taking data from I/O pairwise observations, embedded into a linear causality schema. Despite its success in unraveling the structural and functional properties of neurons, it fails to explain how assemblies of neurons can express cognitive functions. One major consequence of this is that the formation of patterns at the level of populations of neurons, that is, the patterns P_{L_i} at level L_i, though arising from an inferior level in the hierarchy, L_{i-1}, for example single neurons, cannot be explained directly from the immediate inferior level patterns $P_{L_{i-1}}$. While at the micro (neuron) level the causality may be linear, at superior levels it is not and the I–O relationships are not as clearly defined as they are at the microlevel. This idea has a direct methodological implication. The epistemology in the macrolevel is necessarily different than that of the level of single neurons, which has clear-cut components. In sharp contrast to this, in the macrolevel the components are inherently coupled with multiple feedback pathways.

Conceptually, linear causality operates as follows. A stimulus S initiates a chain of events in a linear temporal sequencing. No effect can precede or occur simultaneously with its cause, and at some instant in the chain, each effect becomes a cause. Linear causality can tell us how a reflex response is caused by a conditioned stimulus. But the reflex response is nowhere near being able to explain the causes of perception or any other fairly complex behavior. Linear causality is strictly determinist, A causes B causes, B causes C and so on. For example, a stimulus excites a sensory neuron, that excites the sensory cortex, that transmits to the motor neurons, that finally contracts the muscle.

The philosohical implications of circular causality has been studied by Marleau-Ponty [333]. The action does not follow the sensation, rather both action and sensation are both cause and effect of the "intentional arc". Circular causality expresses the interaction levels in a hierarchy. Brain's hierarchical organisation is endowed with top–down constraints that apply to lower levels of the organisation, which in turn, affects the workings of higher level functions. In biology, it is very rare to find one function that is located in a single component. Rather the function arises from the interaction between several components. A possible exception that confirms the rule, is the haemoglobin that transports oxygen in the red blood cells of vertebrates. The interaction between components and levels in a system is at the core of the bidirectional causation that circular causality describes. Circular causality in brain networks is created by virtue of synaptic interactions between neurons. In linear causality, the stimuli are always injected into a passive brain which has sensors attuned to the stimuli. Contrary to this view, the concept of circular causality is necessary to cope with the self-organising chaotic dynamics within the brain. For a critique of the idea of circular causality see [334].

7.4.2.1 Granger Causality

Granger causality is a term for a specific notion of causality in time-series analysis [335]. Originally, the term was coined by the Nobel laureate economist Clive Granger who used the concept in econometrics. In neuroscience, Granger causality analysis is used to understand the relationship between network structure and network dynamics. Causal network analysis provides a radically new insight into the mechanisms underlying higher cognitive functions, including consciousness. Neural correlations are not explanatory and there is an urgent need for theoretical concepts that connect the phenomenological aspects with neural dynamics. Causal network seems to be useful in revealing the general principles governing causal interactions between long distance neural populations, and also as a framework to test the functional connectivity.

At it stated in [336] "Causality analysis makes no assumption about information processing or neural code." However, Granger causality has been more successful in generating heuristics for the analysis of biological data in bio inspired robotics, for example in the robot Darwin-X, than in help understanding the global functioning of the brain. Some drawbacks of Granger causality are:

linearity, stationarity and dependence on observed variables. Furthermore a causal core analysis requires exhaustive knowledge of structural and functional connectivity which, currently can not be supplied with real data, except for simulated environments. On the positive side, the "causal approach" will strengthen the mesoscopic analysis in brain connectivity and dynamics. This is a necessary step because it has historically received minor attention if we compare it to the microlevel of single neurons, or with the macroscopic level of brain systems.

7.4.3 *Cognits, Neurocognitive Networks and Local Populations*

The study of perception has held as a central position in philosophy, psychology and brain science. A lot is known about how our cortical areas e.g., auditory or visual, transduce and internally represent the external stimuli. These cortical sensory areas are defined based on cytoarchitectural properties. For very simple and primitive features of the external world, the final pattern of neural activity is isomorphic with the perceived reality. In this case, the internal representation of the world is mainly topographic. But the functional properties of neurons loose their specificity as soon as the internal representation is not isomorphic, so the topographic map is superseded by patterns of connectivity able to deal with more abstract features. To sum up, the neural representation of holistic perceptions are embedded in recurrent patterns of connectivity. In order to understand the neural mechanisms in a working brain, we need to challenge the concept of functional representation. An agenda that acknowledges the crucial importance of the representational power of patterns of connectivity and its dynamics must be put in place.

A main premise is that cognitive functions take place within and between neuronal networks, mostly in the cortex. This assumption and how to deal with it, is at the core of the gap between the neural sciences and the cognitive sciences. The mind/brain problem is the philosophical stand for the same gap. The above assumption can turn into a commonplace when not addressed with the appropriate language. In the cognitive sciences, concepts extracted from levels of description as disparate as the mathematical models of activity of the neurons or even deeper, like the molecular composition of neurotransmitters, cohabit with purely linguistic descriptions of emotions or social behavior. Surely, a new language is needed to help bridge that gap.

The concept of meaningful pattern here exposed is germane with the term "cognit" coined by the neurobiologist J. Fuster [337] and with the concept of "neurocognitive networks" conceived by Bressler and Tognoli [320] and Bressler [338]. The term "local population" of neurons in the olfactory bulb studied by W. Freeman [92, 339] are also alike with our meaningful patterns. More precisely, Fuster's cognits are the structural substrate of cognitive operations and represent items of knowledge. Thus, a cognit is a cognitive representation that evolves over time due to the maturational and experience-based process undergone in the brain.

In Bressler's view, in order to understand how large scale neural structures organise to perform cognitive functions, we must be able to first identify the meaningful networks or patterns that express those cognitive functions. Thus, neurocognitive networks are conceived as large-scale systems of distributed and interconnected neuronal populations that mediate in cognitive functions. Freeman's approach is mainly concerned with the introduction of an intermediate level – mesoscopic – in which the dynamics of local neurons is studied in the olfactory system, the simplest and phylogenetically oldest of all mammalian sensory systems. In Freeman's view, the proper element for mesoscopic modeling is the local population rather than the more general concept of network of neurons. Contrary to local populations, network of neurons are not necessarily local. The components in a neural network can be separated by large distances in the brain, for example, networks connecting neurons of different hemispheres or cortico-talamic networks. The locality property utilised in Freeman's mesoscopic models, facilitate their modeling because the dendrites of neighbor neurons, as opposed to the axons whose operation is non linear, can be treated as a linear integrator. So, focusing on locality we steer clear of the non linearity in long distance axons connections. However, the identification of meaningful networks or patterns that mediate in cognitive functions, as promised by the different approaches shown above is daunting. Indeed, when dealing with broadly distributed connections of a large number of components, highly coupled with non linear dynamics, the resulting behavior of the neurocognitive networks are in many cases impossible to control and predict. The problem is as follows: assuming that we know how to describe the dynamics that neurons would exhibit isolated, and assuming that the dynamics of the temporal patterns of those isolated neurons have a well-known long-term behavior, for example, stable fixed points or chaotic attractors, the global behavior is still missing [94].

7.5 Network Based Approach for Brain Connectivity

The visual identification of the overall connectivity patters in functional magnetic resonance imaging (fMRI) has been approached with both parametric and non parametric statistical analysis. In the former, statistical parametric maps of brain activation are built upon voxel-wise analysis location [340, 341]. This approach has been successful in the identification of for example, motor networks, but it shows important limitations when the seed voxel cannot be easily identified. For example, in brain areas with unclear boundaries i.e., cognitive networks involved in language or memory. Principal Component Analysis (PCA) is a data drive or model-free approach able to replicate the functional connectivity maps obtained with the seed voxel approach [342]. However PCA has the important caveat that the time courses of the components are constrained to be orthogonal. Independent Component Analysis (ICA), also a model-free approach, does not constrain the time courses of the components to be mutually orthogonal. ICA has been consistently applied to fMRI data in resting state allowing to separate resting fluctuations from other signal variations, resulting on a collection of spatial maps, one for each independent component, that represent functionally relevant networks in the brain [343–349]. While ICA has the advantage over model-based methods that it is unbiased, that is, it does not need to posit a specific temporal model of correlation between regions of interest, the functional relevance of the different components is, however, computed relative to their resemblance to a number of networks based on criteria that are not easily formalized [350].

More recently, researchers using graph-theory based methods have been able to not only visualize brain networks, but to quantify their topological properties [351]. The seminal work of Erdos and Renyi in Graph theory focused on random graphs. In this approach graphs were generated by some random process [352]. Erdos–Renyi (E–R) proposed a very simple algorithm to build random networks. Each node is connected with another node with probability p, and the distribution of nearest neighbors follows a Poisson distribution. Thus, a random graph is a statistical ensemble of all possible graphs where each member of the ensemble has identical probability of realisation. It ought to be said that "random" refers to the assignment of a probability value to a link between each pair of nodes, and not to lack of structure. Despite its lack of realism – biological networks are neither randomly nor regularly organized – this approach facilitates the study of nonlinear dynamics which is a principal characteristic of empirical networked systems. The idea is to focus on the dynamics of systems coupled in uniform ways, defined in terms of differential equations and recurrent feedback interaction models. The final goal is the identification of universal dynamical laws and specific mechanisms e.g., self-organization, that mediate in progressive coupling and broken symmetry between individual heterogeneous components [353–355].

Graph theory is the mathematical framework for the representation of complex networks. Differential equation models are not well equipped to deal with network phenomena. Graph theory-based approaches model biological systems e.g., the brain, as a complex network represented graphically by nodes connected by edges. However, the E–R model and other similar approaches based on regular or random graphs are not a viable alternative to model biological networks. For example, metabolic networks – nodes denote the metabolites and the edges represent reactions – are far from being a random graph of the Erdos–Renyi type [356, 357]. Furthermore, the distribution of the number of connections is not Poisson but it approximately follows a power law – a small number of nodes (metabolites) participate in a large number of connections (reactions) and an overwhelming number of nodes have one or two connections. Large-scale anatomical connectivity analysis shows that brain topology is neither random nor regular, it is rather characterized by small world architectures – highly clustered nodes connected thorough relatively short paths [358, 359].

A network-based approach able to capture properties related to the network degree distribution, clustering, modularity, hierarchy etc. has notably increase our understanding in the relationship between brain structure and function [360]. The existing bibliography of graph-based network analysis applied to the brain is too large to be described here. The point that needs to be emphasized is that the characterisation of biological networks, initially described with strictly theoretical models, is now addressing empirical biological problems in a systemic way.

7.5.1 Network Analysis for Brain Connectivity

Since the end of the 1990s the study of the physics of complex systems has attracted an important number of researchers of different backgrounds. Network science aims at the characterization, by means of the identification of statistical properties, of the underlying topological organization of complex networks with specific organizational principles that are not typical in random graphs or regular lattices. Network based approaches provide a system level understanding. Specifically, complex network analysis applied to brain connectivity makes available a relatively uncomplicated toolbox to tackle the main principles of structural and functional organization i.e., segregation and integration [361]. Segregation refers to the structural and functional subdivision of the brain into specialized regions [362]. However, sensory perception, cognition and awareness require to integrate information from the distinct brain systems, that is, functional integration [363, 364]. The complementarity of these two features clearly calls for a system or network based approach.

Network analysis requires to have available an adjacency or association matrix which denotes the correlation between the different network components. It is crucial to realise that different connection topologies can generate different modes of neuronal dynamics. The network topology – given by the connectivity matrix – is mute about the neural mechanisms that qualitatively affect the dynamics of the network [365]. For example, the equation of the network dynamics for point-like neurons (dendrites are ignored) requires the two basic ingredients: $v_i(t)$ – the membrane potential of neuron i at time t and $u_i(t)$ – the firing rate of neuron i at time t. It is normally assumed that the firing rate, $u_i(t)$, is a function of the membrane potential $v_i(t)$. Thus, $u_i(t) = F_i v_i(t)$. The dynamics of $v_i(t)$ is given by

$$v_i(t) = \int_{t_0}^{t} G_i(t-s) \sum_j w_{ij} F_j(v_j(s-\delta_{ij}))ds \tag{7.1}$$

where w_{ij} is a scalar that represents the synaptic weight between two neurons i and G a postsynaptic function. The connectivity matrix w defines an effective geometry in which the dynamics of the system evolves, that is to say, it specifies a particular connectivity pattern that might be identified with a network model e.g., small world, but does not say anything about the neural mechanisms that qualitatively affect the dynamics of the network. Network analysis, in particular, graph theory applied to brain networks allows us to quantify topological properties of the network. Network parameters such degree distribution, clustering, characteristic path length, hierarchy, modularity etc. define different network models that can be used to characterize structural and functional connectivity patterns in the brain. A graph is just a graphic representation of the adjacency matrix. Graph theory is agnostic about the kind of brain connectivity defined in the adjacency matrix. Thus, structural, functional or causal connectivity data can be represented as a graph and routinely analyzed with the available network metrics. Brain structural networks may refer to both morphological and anatomical connectivity. The former is built upon statistical correlations between morphological descriptors e.g., gray matter or cortical thickness, and the last typically refers to fiber connections using diffusion MRI data. Brain functional networks define temporal correlations between neurons or regions in the brain. There are multiple alternatives to estimate the functional connectivity between brain areas, including wavelet correlation [366], Pearson correlation [367] and mutual information [368].

A network is defined in graph theory as a collection of nodes N and a collection of edges E, $G = (N, E)$ that connect pairs of nodes. Multiple choices are available for characterizing the topology of graphs. Network properties also called network metrics can be roughly classified into global and local. The most common global properties include among others degree distribution, clustering coefficient, the characteristic path length and hierarchy. Local properties are essentially *network motifs* i.e., subgraphs that are significantly overrepresented.

The degree of a node is the number of edges linked to it and therefore measures the connectivity of a node with the rest of nodes in the network. The degree distribution $P(k)$ is the probability that a node has k neighbors. Most real networks deviate from the Poisson degree distribution of random networks, often with long tails towards high degrees. Networks with a degree distribution that follows a power law, $P(k) = k^{-\gamma}$ with $\gamma = [2.1, 2.4]$, are called scale free networks. Scale free networks are fractal in the sense that no single characteristic scale can be defined [369]. They have structural self-similarity, or borrowing the expression from modern physics, scale free networks present symmetry under scale change. This kind of network is resilient to random failure but vulnerable to targeted attacks i.e., injury of nodes with high connectivity degree. It is of note that the use of power law in modeling experimental data has been criticized. For example, in relation with misleading statistical estimation due to insufficient data [370] and the poor explanatory insights of power laws [371].

The clustering coefficient measures the tendency of the network to form highly interconnected regions called clusters or cliques. The clustering coefficient $\sigma(n)$ of a node n in a network, is defined as the probability that two nodes m and r which are connected to the node n are themselves connected, that is, the probability of having a "transitive triple" [372]. The average of $C(n)$ over all nodes in a network is the clustering coefficient σ of the network. It is worth of note that the clustering of a node, $\sigma(n)$, is a local measure of the efficiency of information transfer.

Path length is the minimum number of edges that must be traversed to go from one node to another. The average path length λ is the average of the shortest path lengths between any pair of nodes in the network, and it measures the overall communication efficiency of a network. Numerically, the efficiency is calculated as the inverse of the average path length [373]. Thus, a network is efficient if the average minimum distance between nodes is small.

In a hierarchical network, nodes with high degree are connected to nodes that are not otherwise connected to each other. This configuration favors top down relationships and is vulnerable to injuries of hubs. The hierarchical structure in a network, β, measures the relationship between the clustering coefficient of the vertices and the degree, $\sigma = k^{-\beta}$. Networks with higher hierarchy values have large positive β values. For example, functional brain networks in children showed significantly lower levels of hierarchical organization compared to adults [374].

7.5.2 Small World Network

With the network parameters σ and λ is possible to determine whether a network has a small world topology. Researchers have detected small work networks in an impressive number of both natural and man made systems, ranging from social acquaintances [375] and the Internet [376] to neural connectivity of the nematode Caenorhabditis Elegans [144] and associations of cortical thickness among regions [377, 378] to cite a few. The small-world model in complex brain networks provides a quantitative framework for the segregation/integration problem and may explain also the efficiency of information transfer at a relatively low wiring cost [146, 379, 380].

Small-world networks have shorter characteristic path length – λ – than regular networks and greater local interconnectivity – σ – than random networks (Eq. 7.2). Put in another way, most nodes do not need to be neighbors of one another to be connected, they still can be reached from every other by a small number of steps.

$$\sigma_{small\,world} > \sigma_{random}\;,\;\lambda_{small\,world} < \lambda_{regular} \tag{7.2}$$

Small world graphs characteristics – high clustering and short path distances – have been reported in cortical structural connectivity [381]. Axons projecting long distances are few in number compared to local axons forming clusters of neurons. An implication of the small world network is that they provide high speed communication channels between distant parts of the brain, facilitating dynamic processes that require global cooperation between neurons, for example in neural synchronisation [382]. Small world topology in mammal brain functional networks has been identified in for example, [383] In addition to this, disruptions in the small world organization can give clues about normal development and pathological conditions. For example, Supekar and colleagues [384] have shown that the deterioration of small world properties such as the lowering of the cluster coefficient, affect local network connectivity, which in turn may work as a network biomarker for Alzheimer's disease. Abnormalities in small-worldness may also have a significant positive correlation in, for example, schizophrenia [385] and epilepsy [386].

Space limitations prevents from a more exhaustive description of network metrics. For an in depth survey of other important network parameters including modularity, betweenness centrality and spectral and community identification measurements etc., see [387]. It ought to be remarked that the different network properties available are not a complete set of the network metrics required to characterize network topology and the complex dynamics that emerge from it. New metrics that foster our understanding of the relationship between topology and dynamics and help us explain network robustness and adaptation are still required [388–391].

7.5.3 *Challenges in Network Based Approaches*

The notable success in the identification of brain systems using network models e.g., small world, may produce the impression that the anatomical and functional patterns are fully understood. Far from that, the discovery of small world networks in the brain is only a very preliminary step towards a comprehensive understanding of how these networks are organized [392]. While network topology constrains network dynamics, this in turn shapes network topology, so the identification of network metrics per se is ineffectual for a fully understanding of the interplay between network topology and dynamics.

The election of nodes and edges in the network may have a significant influence in the representation of the connectivity map. In the absence of a gold standard for the definition of nodes and edges, the reproducibility and stability of results might be affected by the choices researchers made in both designing the experiment and the analysis. For example, network nodes can be brain regions according to a prior brain atlas or voxels, and the specification of the network edges is subject to variations depending on the threshold and the correlation analysis adopted. While high reproducibility in small world has been recently reported in vivo diffusion tensor imaging (DTI) data [393], we must always have present that the brain is a complex system operating at different spatiotemporal scales. Deuker et al. [394] have shown that reliability of graph metrics varies considerably over different frequency bands, improving in the lower frequency. Hayasaka and Laurienti [395] found that small-worldness varies depending on the granularity in the definition of nodes i.e., small world is more robust in higher resolution networks (circa 160,000 voxel like nodes) compared with networks with 90 nodes representing region areas. Thus, additional work in contrasting multi-center and longitudinal datasets is required to assess the reproducibility and stability of functional connectivity networks.

Arguably, the small world architecture – short path length and high clustering – is an archetypal brain network. While network-based studies have been successful in delineating generic network properties, such as path length or clustering, additional work is still needed. Robustness in brain connectivity has been typically approached in terms of the impact that the complete disruption and/or removal of a network component has in the network topology [396]. However, by focusing on the topology of the network, factors that may play a key role in the network's vulnerability to failures can be neglected. For example, it has been suggested that patients with Alzheimer's disease show an increment in brain activity in certain areas relative to healthy subjects that, arguably, compensates for the disease related atrophy [384]. Recent studies have shown the potential of network analysis in neurodegenerative brain diseases [385, 397, 398]. Neurological disorders target large scale functional and structural networks rather than specific loci or sub-networks. The study of aspects relative to network vulnerability, specifically its robustness, may help in the discovery of network biomarkers for early diagnosis of neurological diseases such as Alzheimer's disease. However, a systemic study of the network topology and its dynamics at individual basis has not been undertaken yet. New tools to compare individual brain networks against a series of network models that refer to either normal development and aging, will foster our understanding of brain functioning, opening new avenues for personalized theranostics (therapeutics + diagnosis).

Network motifs can be regarded as the building blocks of the network and therefore are likely to play a critical role in the network dynamics [399, 400]. However, the identification of a statistically significant motif does not automatically imply that it has biological relevance. Furthermore, the comparison of motifs inside a large graph is computationally infeasible due to the NP-completeness[2] of the underlying subgraph isomorphism problem [401].

How brain structural connectivity affects and is affected by the network dynamics that is at the basis of cognition, is far from clear. Network theory allows the analysis of connectivity patterns extracted from both structural and functional data, gather with different imaging techniques. Combined structure-functional analysis is particularly challenging due to the very different nature of the data sets e.g., DTI and fMRI. Hagmann and colleagues have been able to show that structural patterns estimated from diffusion imaging and functional patterns in resting state fMRI (RS-fMRI) are significantly correlated [402]. However, there is a caveat in this approach. Structural connections are good predictors of functional connections in resting state experiments i.e., neurovascular activity measured over minutes with no task. But, the likelihood is that for shorter period of time and task based experiments, the correlation between structural and functional connectivity will decline.

7.6 A Categorical Framework for Network Theory

The explosion in biological and other real-world network data, due to either the improvement in the technological capability of measuring spatiotemporal patterns, and the increasing attention that network-based approaches from researchers with

[2]NP refers to nondeterministic polynomial time and NP-complete is a subset of NP. Computational scientists are still debating whether NP-complete problems are solvable in polynomial time.

very different backgrounds, is challenging our capacity to represent, analyze and control the dynamics of complex networks. With the availability of whole-genome sequences, geneticists are faced with the challenge of decoding the information contained within these DNA sequences. By analogy, comparing large cellular e.g., neural networks, if done with the best possible methodologies, will revolutionize our understanding of brain function. The bottleneck, however, might not reside in resolution measurements but in our ability to build tools and theories for large network analyses.

Algorithms to detect network motifs i.e., small over-represented subgraphs within a graph, rely on a two step process. First, calculate the number of occurrences of a sub-graph against a null-model, which is defined by a large ensemble of randomized networks with the same degree for each node, and second, evaluate the significance of each prospective motif. The number of the motif appearances is calculated through an isomorphism between the motif and a repertoire of network motifs which are assumed to have particular functional importance. For example, one of the most prominent network motifs in the E. coli is the feedforward loop which accelerates or delays responses to stimulus steps. In the brain, feedback loops for both recurrent excitation and inhibition are recognisable motifs. Thus, the identification of motifs relies upon finding isomorphisms between subgraphs. This approach has been criticized in light of the questionable biological relevance that statistically significant sub-graphs may have. Recent studies have shown that when the network context is taking into account, alternative explanations in terms of functional uses of motifs can be easily found [403–405]. Furthermore, the detection of network motifs is computationally expensive. The subgraph isomorphism problem is P-complete i.e., the time required to solve the problem increases very quickly as the size of the problem grows. By the same token, to compare two large networks, rather than small subgraphs (motifs), may be computationally inaccessible. It is however of note that the subgraph isomorphism problem can be solved in linear time for simple planar graphs i.e., the number of edges is at most O(n) and it can be drawn in such a way that no edges cross each other.

7.6.1 Exploring Motifs with Graph Homomorphisms

The standard approach used for structural network analysis is based on finding isomorphisms between, for example, a subgraph within a graph. When the subgraph isomorphism problem is computationally infeasible, network comparison can be investigated heuristically i.e., network properties as degree, clustering, centrality etc. Here, we propose an alternative approach to investigate graph structure, in which, rather than looking for isomorphisms between graphs we study the homomorphism category of a family of graphs where each graph is understood by means of its relationships with other graphs in the category. It ought to be remarked that while networks define binary relations, categories can deal with relational systems with different arities, that is binary, ternary etc. The main rationale is to show that category theory provides the means to overcome the limitations of graphs while at the same time including graph theoretic tools. This categorical framework applied to graph theory allows us to set the problem of graph comparison in a more general way, and at the same time, continue using graph theoretic methods.

A category whose objects are graphs and the arrows are homomorphisms between graphs, or increasing the complexity of the objects, we can compare different categories using functors or compare functors by natural transformations. By defining homomorphisms between graphs we are imposing an algebraic structure to an initial set of graphs that is deprived of this categorical outlook.

Homomorphisms can be propagated using the composition operator i.e., the composition of two homomorphisms is also a homomorphism. It ought to be remarked that contrary to endomorphisms which are always composable, two homomorphisms can be composed only if the codomain of the first one is the domain of the second one. Homomorphisms of graphs preserve adjacency i.e., for any two graphs G_1, G_2, a homomorphism $h : G_1 \to G_2$ is a mapping that relates the vertices of graphs G_1 and G_2 in such a way that if vertices ab are connected in G_1 then $h(a)h(b)$ are connected in G_2. We say that G_1 is homomorphic to G_2 if there exists a homomorphism $h : G_1 \to G_2$. An interesting property of graph homomorphisms is that they do not increase distances. For example, homomorphisms from G_1 to G_2 map paths in G_1 to walks in G_2. Formally, given the homomorphism $h : G_1 \to G_2$, $d_{G_2}(h(a), h(b)) \leq d_{G_1}(a, b)$, for any two vertices $a, b \in G$ and $d_G(a, b)$ being the length of the shortest path between a and b in G. The demonstration is trivial since every walk from $h(a)$ to $h(b)$ of length k contains at least a path from those two vertices, $d_{G_2}(h(a), h(b)) \leq k$ [406].

Now, it is possible to study in all its richness the (algebraic) structure of graphs denoting the relations between them, within the homomorphism category of graphs, with respect to the composition operator. The category of graphs is composed of categories and homomorphisms between categories. While an isomorphism between two objects, for example two graphs G_1, G_2, is a bijection between the set of nodes, such that any two vertices that are adjacent in G_1 are also adjacent in G_2, a graph homomorphism is a mapping between two graphs (not a bijection) that respects their structure. Homomorphisms preserve all the relations that exist in a relational system, e.g., a graph. The homomorphisms among a set of graphs impose the algebraic structure of a category. Thus, the spirit of this method is to study the relation between graphs in the given category, via the agreed morphisms, in this case homomorphisms of graphs.

Graph homomorphisms have been used in statistical mechanics, in particular in counting Widom-Rowlinson configurations in graphs, which is a particular model of a gas [407, 408]. Some counting problems can be solved by counting the number of homomorphisms from the graph of interest G_1 to a given fixed graph G_2. Here, a configuration is to assign states to the sites of a structure according to the specifications defined by a particular model e.g., the Widom-Rowlinson gas model of n-particles, It may be possible to build on this idea to use graph homomorphisms to delineate graph relationships of connectivity maps. This may open new avenues to address new questions and to restate old ones within a new categorical framework. For example, a very promising strategy would be to investigate the network degeneration hypothesis – neurological diseases target functional neural networks modifying its topological properties [397] – within the context of the homomorphisms category of graphs. By the same token, the study of graph homomorphisms in brain connectivity patterns may provide an unifying perspective to the study of the causes and prognosis of the approximately 600 known brain disorders.

7.6.2 *Category of Paths*

Category theory has been successfully applied to qualitative analysis, mainly in theoretical computer science to deal with programming language semantics. Nevertheless, the potential of category theoretic tools for quantitative analysis of networks has not been tackled so far. Statistical methods to investigate graph structure typically rely on network parameters. Category theory can be seen as an abstraction of graph theory. Thus, new categorical properties can be added into network analysis and graph theoretic constructs can be accordingly extended in more fundamental basis. This is however, terra incognita for network theorists. Here we propose to carry out network quantitative analysis by defining network metrics inspired in the theory of categories. Before introducing the methodology to study networks properties such as clustering, via the theory of categories, we need to understand three points. First, the conditions by which a graph G underlies a category C, second, a graph can generate a free category, namely the category of paths C(G) and third, a category C determines an underlying graph UC.

First. For a graph to underlie a category it needs that both identity and composition law hold. Figure 7.4 shows examples of graphs and whether they underlie or not a category. It might be emphasized that the structure of a category is richer than the structure of a graph in the sense that a category can be seen as a graph that hold certain axioms. Thus, only some graphs are also categories whose objects are the vertices of the graph and the arrows the graph edges, but every finite category can be represented by a set of graphs.

Second. Graphs and categories are both usually represented as diagrams of vertices and arrows, with the difference that in graphs, contrary to categories, composite arrows are not provided. Identity arrows are also usually omitted. Graphs are also called "precategories". This is because any graph $G = \{A \rightrightarrows O\}$, where A is a set of arrows and O a set of objects related through the domain and the codomain functions, can be used to generate a category of paths $C(G)$, with the same set of objects O and as arrows the paths defined by the composite edges of G. The category of paths $C(G)$ generated in this way is also called the Free Category

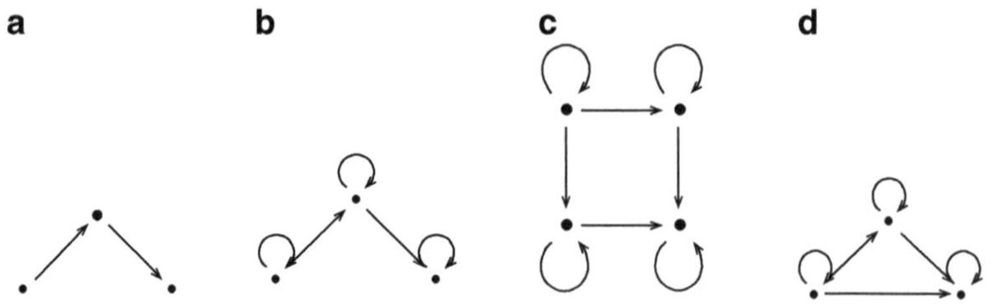

Fig. 7.4 Each category is a directed graph but the converse does not hold because in graphs the notions of internal rule – composition – and identity are lacking. (**a**) A graph that does not underlie a category since the composite among the *arrows* is missing and there are no identity *arrows*. (**b**) A graph that does not underlie a category since the *arrows* are not composable – the codomain of one *arrow* is the domain of the other *arrow* – and the composite is missing. (**c**) A graph that does not underlie a category since the diagonal composites are missing. (**d**) A graph that does underlie a category since the composite and the identities are present

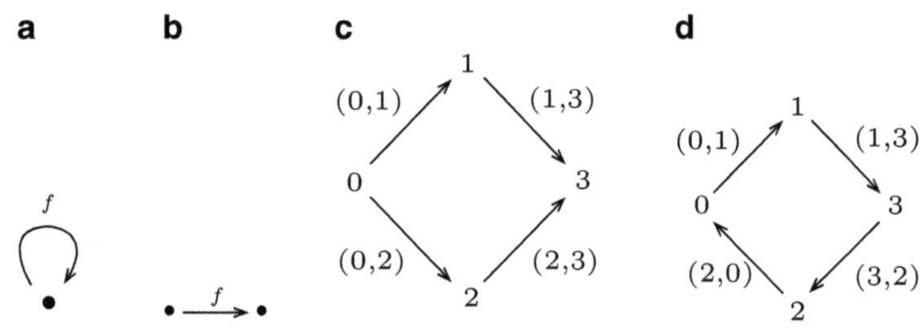

Fig. 7.5 A graph G, with a set of objects, O, and set of *arrows*, E, generates the free category C(G) on the same set of objects, and the *arrows* given by the paths of composable edges of E

generated by the graph G. Note carefully that if G is already a category i.e., identity and composition hold, the generated category is not just the diagram depicted by the nodes and arrows in G. The generated category of paths C(G) is always bigger.

Proposition 7.1. *Let G be a directed graph, then C(G) is the category of paths through G, where the objects of C(G) are the nodes of G and the arrows of C(G) are the paths through G.*

The demonstration is as follows. Composition associativity in C(G) holds because concatenating paths end to end is associative. Each object has an identity arrow on C(G) because each node in G has an associated path of length zero. Thus, the identity arrows are the paths of length zero. Let us see this with an example.

The graph Fig. 7.5a consisting of a single arrow f, generates the free category where all the arrows are produced by the iterative composition of f i.e., $1, f, f^2 \ldots, f^n$. The graph in Fig. 7.5b represents one arrow connecting two different vertices and generates the free category whose objects are those same vertices, and the arrows are given by f plus two identity arrows. The graph in

Fig. 7.5c of four vertices and four edges, generates the free category whose objects are those same vertices and 11 arrows for each possible path i.e., 4 paths of length 0 (identities), 4 paths of length 1 depicted in the graph, plus the two formal composites $(3, 1)\circ(1, 0)$ and $(3, 2)\circ(2, 1)$. The graph in Fig. 7.5d which represents a loop of four vertices connected through four edges, generates the free category of four objects and infinitely many arrows.

Third. Every category C determines an underlying graph UC, where the vertices of the graph UC are the objects in C, and the edges are the arrows forgetting which are composite and which are identities. Every functor between two categories $F : C \rightarrow C'$ is also a morphism of the underlying graphs $UF : UC \rightarrow UC'$. Note that the forgetful functor from categories to graphs, $F : Cat \rightarrow Grph$, works as a generalization the previous statements.

To summarize, each category is a directed graph, but the contrary does not hold because graphs have no composition. However, it is always possible to convert each directed graph into a category – the free category or category of paths.

7.6.3 A Methodology to Study Network Topology via Categories

Here we elaborate on the previous discussion to provide a methodology to study graph characteristics via the theory of categories. More specifically, network properties such as clustering and modularity are here modeled in quantitative basis with colimits and limits respectively.

Given a graph G, it is possible to calculate the clustering coefficient of G by calculating the number of colimits in the category of paths C = C(G), on the same set of vertices of G and the arrows of C given by the paths of composable edges in G. This procedure is schematically depicted in Fig. 7.6. In the first step, we calculate the free category C = C(G) generated from the initial graph G (the objects of C(G) are the vertices of G and the arrows the composable edges of G), to then calculate the number of colimits in C. Observe that Step 1 becomes in the case in which G is already a category redundant i.e., we do not need to generate the free category of G when G already underlies a category. In step 2, the underlying graph UC of C is specified (the vertices of UC are the objects of C(G) are the edges of UC are the arrows of C(G) forgetting which arrows are composite and which are identities). In the third step, the input is the underlying graph of the free category of the graph under study (note that this transformation between the initial graph G and the associated graph UC via the category C(G) is necessary in order to calculate the number of colimits of C(G)) and the output is n different graphs $\{G', G'' \dots G^n\}$ produced according to a specific model. A number of network models have been applied to study the topological properties of complex networks. The election of one model or other will depend on the specific features in which we are interested in. For example, given a graph G we can compare it against graphs with high clustering by rewiring each edge of G with probability $p \sim 0$. To maximize the number of loops in

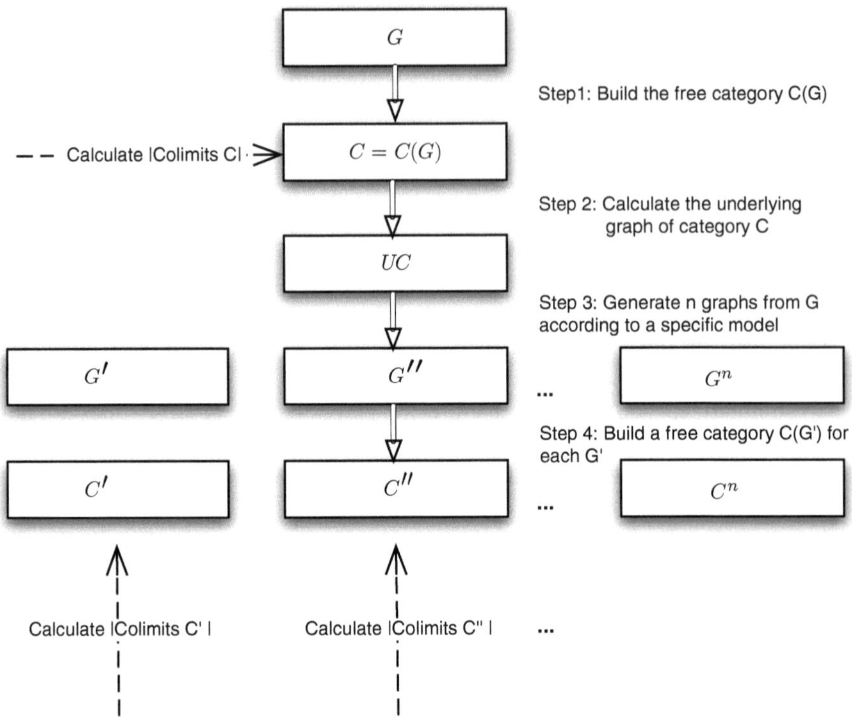

Fig. 7.6 The figure depicts the study of network properties, specifically clustering, through colimits. In Step 1, the initial graph G generates the free category or category of paths $C = C(G)$ having as objects the vertices in G and as *arrows* the paths of composable edges in G. With the category C in place, we calculate the number of colimits, cl_C. The clustering of G is calculated through the position of cl_C in the distribution of colimits given by the categories $\{C(G'), C(G'') \dots C(G^n)\}$ generated from the graphs $\{G', G'' \dots G^n\}$ according to a specific model e.g., rewiring each edge with probability p. Thus, in Step 2 we calculate the underlying graph UC of category C, to then in Step 3 generate n graphs $\{C(G'), C(G'') \dots C(G^n)\}$. Finally in Step 4, we build the category of paths for each of the randomly generated graphs $\{C', C'' \dots C^n\}$ to allow us calculate the number of colimits for each category, $cl'_C, \dots, cl^{n'}_C$. The clustering coefficient of G is computed as a relative measure of each observation i.e., the number of colimits in C(G), cl_C, from the reference sample given by the vector of the number of colimits for each category, $cl'_C, \dots, cl^{n'}_C$

the generated graphs, p needs to be close to 0. This is because when $p = 0$ we have an ordered lattice with a high number of loops, while in the other extreme, $p = 1$, the network becomes a random graph with a few loops. This idea is elaborated in the small-word model proposed by Watts and Strogatz [144] and Strogatz [409]. Other algorithms to generate graphs with small-worldness characteristics, using methods other than rewiring edges have been proposed [410]. In Step 3, once the n randomly generated graphs have been produced, we generate for each of them the free category or category of paths $\{C(G'), C(G'') \dots C(G^n)\}$. And finally in

Step 4, the number of colimits for every category of paths allow us to calculate the clustering coefficient of the initial graph G. Note that the problem of characterizing the types of diagrams that admit a colimit in a free category has been explored in [411]. However, a computational methods to compute colimits in a free category i.e., category of paths, does not exist.

7.6.4 Quantitative Analysis of Clustering and Modularity with Colimits and Limits

In what follows a quantitative network based analysis of the categorical constructs limit and colimit is provided. For a definition of these concepts see Chap. 3. We must remind that intuitively, a limit can be seen as a collection of elements with compatible features for all of them. A colimit is the dual of a limit and works as an integrator of informational units within a category. Thus, limits "lay out' parts and colimits "glue" parts together.

Two alternative definitions of clustering and modularity based on universal constructs of category theory are described. First, a measure of clustering and modularity relative to the entire network and second, a new approach based on local clustering and modularity coefficients, that is, relative to each node. In the first definition of clustering (and modularity) the clustering coefficient of the initial graph G is calculated as a relative measure of the number of colimits in C(G), cl_C, from the reference sample given by the vector of the number of colimits for each category, $cl'_C, \ldots, cl^{n'}_C$. Thus, the clustering coefficient of G represents a distance of the n sample points from a common point. The problem of calculating the clustering coefficient becomes the problem of measuring the "nearness" of the observation, $Y = cl_C$, to the data distribution, $X = cl'_C, \ldots, cl^{n'}_C$. As X is univariate data, the distance can be expressed in terms of the standard deviation. When X follows a normal distribution the distance from the mean can be easily computed with the z-score value. Thus, for an observation y, the z-score of x is $z = \frac{x-\mu}{\sigma}$, where μ is the mean and σ is the standard deviation. The z-score is a dimensionless quantity that tell us the number of standard deviations that the observation y is from the mean.

The clustering coefficient is defined by the z-score and is defined within a range $z_{p=0}$ and $z_{p=1}$, where $z_{p=0}$ is the distance to a data population given by the number of colimits in ordered lattices (a high number of loops) and $z_{p=1}$ is the distance to a data population that represents the number of colimits in random graph (a few loops). Thus, the clustering coefficient of graph G is the distance from the number of colimits of C(G) to the data distribution given by the number of colimits of n generated categories according to a specific model. For normally distributed data, the clustering coefficient is given by the z-score.

The distance can be generalized to multivariate normal distribution i.e., the clustering coefficient has more than one variable, for example the number of colimits, the number of pushouts etc. The Mahalanobis distance can be used to calculate such a distance [412]. This distance is often used in the context of outlier detection in multivariate data.

$$d(Y, X) = \sum_{1}^{N} (y - \mu), \frac{(y - \mu[X])}{(\sigma[X])} \tag{7.3}$$

where μ and σ are the sample mean and covariance matrix of the data in X i.e., the number of colimits for each category, $cl'_C, \ldots, cl^{n'}_C$, and Y is the observation or number of colimits of the underlying category of the initial graph i.e., cl_C. Note that the Mahalanobis distance reduces to the Euclidean distance when the covariance matrix is the identity matrix $\sigma[X] = I$

By the same token, the modularity coefficient can be computed using the number of limits. Thus, the modularity coefficient is computed as the distance of the number of limits of C(G) to the data distribution given by the number of limits of n generated categories according to a specific model.

An alternative approach is to calculate the clustering coefficient of each node and average them to obtain the network clustering coefficient. We can calculate the modularity and the clustering coefficient for a node i in a graph G as a function of the number of pullbacks or pushouts respectively, relative to the number of triples. A triple is a set of three vertices where each vertex can be reached from each other (directly or indirectly), i.e. there must exist a central vertex adjacent to the other two vertices. The universal properties product, co-product, pullback and pushout are all specific instances of the more general notions of limits and colimits. Thus, the modularity of vertex i, M(i), can be formulated as:

$$M(i) = \frac{N_{pullback(l)}}{N_3(i)}, \tag{7.4}$$

where $N_{pullback(l)}$ is the number of vertices $l, l \neq i$ such that l is the pullback of the vertices j, k along the central vertex i and $N_3(i)$ refers to the triple shown below.

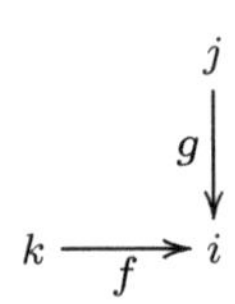

The modularity coefficient for a given graph G calculated via the average of the nodal modularity M(i),

$$M(G) = \frac{\sum M(i)}{n} \tag{7.5}$$

Modularity entails a sensible decomposition of the given graph into communities i.e., sub-graphs where all of its vertices have more connections between them than with the rest of the network. Despite the importance of the concept of modularity in complex networks, a universal definition of network modularity is still lacking. For a review on the problem of community identification, see [413] and [414].

The clustering coefficient of vertex i, C(i) is given by the identity:

$$C(i) = \frac{N_{pushout(l)}}{N_3(i)} \tag{7.6}$$

where $N_{pushout(l)}$ is the number of vertices $l, l \neq i$ such that l is the pushout of vertices j, k along the central vertex i, $N_3(i)$ refers to the triple shown below:

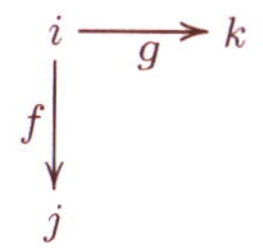

The clustering coefficient of graph G based on the nodal clustering C(i) is the average of all the clustering of the vertices.

$$C(G) = \frac{\sum C(i)}{n} \tag{7.7}$$

Other definitions of clustering coefficient are also possible and compatible with the categorical approach proposed here. For example, clustering can be defined with intensity of the triangle subgraphs [415]

$$C(i) = \frac{2}{k_i(k_{i-1})} \sum_{(j,k)} (\hat{w}_{ij} \hat{w}_{jk} \hat{w}_{ki})^{1/3} \tag{7.8}$$

where weights are scaled by the largest weight in the network, $\hat{w} = \frac{w_{ij}}{\max(w_{ij})}$. The clustering coefficient of the graph can be now be calculated as a function of the degree k.

$$C(k) = \frac{\sum_i C(i)\delta_{k,k}}{\sum_i \delta_{k,k}} \tag{7.9}$$

where $\delta_{i,j}$ is the Kronecker delta, therefore $\delta_{k,k} = 1$.

7.6.5 Seeing Small World and Scale Free Networks with Categorical Lens

To summarize, category theory, and in particular, "universal constructions", such as the colimit, effectively models the "glueing" of a pattern of linked objects in a graph. By generalizing networks using category theory we can address questions and elaborate answers in a more fundamental way without waiving graph theoretic tools.

Here we provide some examples on how the methodology described in Sect. 7.6.3 can shed light in the study of complex network metrics.

The small world property – most vertices can be reached from the others in a small number of steps (short paths) – is found also in the Erdös-Rényi random model. However, a most interesting property of small world networks is that of large number of loops of size three i.e., if vertex i is connected to vertices j and k, there is a high probability of vertices j and k being connected. Loops of size three (Δ) are commutative triangles in category theory. There is abundant literature that show that small world and scale free properties characterize an important number of real complex networks. But these two attributes may not be compatible. The small world property can be mathematically defined as the average diameter of the network, l, increases logarithmically (slow) with the total number of nodes, N, $l =\approx lnN$ or

$$N = e^l \tag{7.10}$$

While small world networks are associated with random patterns of connections with a characteristic degree, scale free networks, on the other hand, follow a power law degree distribution in which some vertices – hubs – are highly connected, while others have only a few connections. Thus, the number of links per edge can be represented by a power law $P(k) = k^{-\gamma}$ with $2 < \gamma < 3$. Scale free invariance and the exponential increase in small world networks (7.10) seems two conflicting behaviors that may need a more fundamental framework. Furthermore, the self-similar exponent that quantifies scale invariance may result far from trivial to calculate in networks with power law degree distributions. The study of connectivity patterns between communities on different length scales may shed some light in the underlying mechanisms that mediate in either small world and scale free properties. Song et al. [416] have shown, through the use of the fractal dimension in scale-free renormalization procedure, that real complex networks may consist of multi-scale self-repeating patterns. They have also being able to reformulate the degree exponent in scale free networks, γ, in terms of two indexes. A similar approach using categorical colimit is also possible. The number of colimits necessary to cover a network and the size of the diagrams with a colimit can be used to define new and more fundamental self-similarity exponents.

To summarize the methodology presented here describes for the first time a computational approach to study network properties via categories, specifically using the number of limits and its dual, colimits, in the underlying category of graphs. The method is general and can be applied to any graph, that is, the graph under study does not need to underlie a category. This would drastically reduced the applicability of the theory. There is a final caveat to be considered. The computational procedure to calculate the number of colimits in a category is not being provided here. The vital issue is to establish a new framework for quantitative analysis of networks using the theory of categories, in which computational neuroscientists and network theorists may tackle in more efficient ways the dynamics of brain cognitive networks.

Chapter 8
A Theory of Hippocampus Structure and Function Based on Category Theory

8.1 Introduction

This chapter provides a theory of hippocampus structure and function based on category theory. In particular, it demonstrates that the co-operation of the grid fields gives rise to a "colimit" which is a place field. Section 8.2 provides an introduction to different theories of spatial representation. Sections 8.3 and 8.4 focus on the brain hippocampus as a representational device, studying its anatomy and physiology, reviewing cognitive theories associated with the hippocampal system. Sections 8.5–8.7 describe place cells, grid cells and head direction cells respectively. Section 8.8 describes a theory of brain spatial representation based on category theory. And in the last Sect. 8.9, a new framework for place and grid cells is described.

8.2 The Nature of Spatial Representation

Prior to get into the biological basis of space and the major role of the hippocampus in spatial representation, I will give a brief keynotes on the different conceptions of space in modern science (since Galileo). Those indifferent to philosophical and theoretical debates may skip this heading and pass directly to Sect. 8.3.

In physics, the main concern is with the properties of the physical space, which is mean to be that inhabited by things, belonging to the external world and independent of the minds that perceive them. A time-honored disquisition in natural philosophy[1] is the absoluteness or relativeness of the physical space. While there is a common agreement in that the psychological space necessarily needs to match aspects of the physical external universe, the disagreements begin when one asks about the nature of space.

[1]Contemporary physics is the modern version of natural philosophy.

J. Gómez-Ramirez, *A New Foundation for Representation in Cognitive and Brain Science*, Springer Series in Cognitive and Neural Systems 7, DOI 10.1007/978-94-007-7738-5_8,

Newton built his laws around the metaphysical concept of absolute space as a "thing" with no relation to anything external to it and inaccessible to the senses. In Newton's dynamics, the concept of absolute motion leads naturally to that of absolute space, the last must exist because the former is real. For example, the centrifugal force creates an centrifugal motion in a body which is an absolute motion that can only be understood assuming that it occurs within an absolute space. Even though we can not experiment the absolute space because is inaccessible to our mind, we can measure the effects of the bodies embedded in such a space. Leibniz is the champion of the notion of relative space which is, of course, neatly at odds with Newton's absolute space. For Leibniz, space cannot exist in the absence of objects, specifically monads which are a very special kind of object conceived by Leibniz – a sort of metaphysical object – with no mass nor extension. The main characteristic of monads is their continuous changing internal state. In Leibniz, the space arises from the relationship between monads. Thus, space is the whole collection of places. In Kant, the conception of space as a "synthetic a priori" lays, on the one hand, on Newton's absolute (physical) space and, on the other, on Euler's hypothesis of space as purely psychological but not derived from sensations of the external world. In this view space is pure intuition that is given prior to all other perceptions, and contains in itself the principles that relate the perceived objects [417]. Now is time to turn to those principles.

In Euclidean geometry we are concerned with three type of objects: points, lines and planes which are not defined and neither can be discovered by repetitive acts, rather they are intuitively given. In this rubric our minds are not only hardwired with these three notions also the basic relations between them are also given [418]. After the discoveries of new geometries in the nineteenth century by Riemann, Lobachetsky and others, the Kantian assumption that Euclidean geometry necessarily reflected the structure of the physical world was no longer tenable. However, it might be said that the existence of Riemann geometry does not invalid the idea that the spatial framework is innately given, as Kant argued, rather it is the metric of the spatial framework that must be revised. A remarkably usual argument against Kant a priories – space, time – is that there are other geometries than Euclidean geometry like for example Riemann. Kant, was a son of his time and could not possibly know about the future theoretical developments in mathematics and physics, of course Kant could never guess what Riemann's mind would fabricate! The problem with this criticism, as I see it, is that there is the expectation that the theory postulated by Euclides – axioms and theories – must be implemented in the brain to validate Kant's view. It goes without saying that axioms can hardly be hardwired in the brain, but, and I think this is where Kant is correct, the concept of distance or metric is actually embedded in our neuronal tissue as the grid cells show.

Poincaré's model of space states that the geometrical space is a matter of convenience, we choose one geometry or other depending on our needs. For Poincaré, space is representative and introspective, and therefore lacks the neutral qualities of the Kantinian space: infinite, continuous, homogeneous and isotropic.

Poincaré's conception of space is thus, egocentric i.e., subject driven. Hull used the "egocentric" conception of space to study spatial behavior in rats. Hull coined the term "habit-family hierarchy" which is an application of mathematical group theory familiar to Poincaré's displacement group – the group of the different movement patterns that can bring the system, for example a rat, to a previous location [419].

Gestalt theory borrowed terms from relativity physics, notably the concept of field which is an overload term. Neural processes conform here a psychological field. The Gestalt school is essentially holistic i.e., the individual parts of the psychological field lack phenomenological meaning, and the entire pattern or field is isomorphic to a percept. Gestalt theory of perception has been criticised for being speculative or descriptive rather than explanatory. Gestalists reflected the concept of an a priori objective space and overemphasized physiological fields but they failed to provide a convincing analysis of behaviour in space according to Gestalt principles. The discomfort of computational neuroscientists with Gestalt Theory is attributable to its inability to cope with neither objective spatial representation nor learning [420].

8.3 The Hippocampus as a Representational Device

How does the mind represent physical space? This is a question that has kept philosophers busy for centuries. In 1971, the philosophical discussions about space representation acquired a extremely powerful and fresh insight when O'Keefe and colleagues, discovered the place cells in the hippocampus of the rat [421].

The experimental study of spatial representation in the brain has since then exploded. The 1970s was the decade of the place cells, neurons that discharge when the rat is in a particular position. In the 1980s head direction cells, neurons that discharge significantly whenever the rat's head changes direction, acquired the attention of scholars. Since 2005 we have been in the era of the grid cells. These discoveries are of major importance in different research fields. Indeed the theory of the cognitive map [420] is rooted in the discovery of place cells in the hippocampus. One derivative of this theory is the map-based navigation capability, that some animals have, and that engineers have been able to replicate in robots [422].

The debate around whether the brain generates a map-like structure or not, seems to have shifted in favour of those who support the cognitive map theory. Indeed, the discovery of place cells, head cells and grid cells suggest so. Yet the underlying nature of the cognitive map remains elusive. Is the representation purely metrical or is topological? Are the maps constructed in the hippocampus built without paying attention to the features of the environment – i.e., metrical maps – or do they reflect the relationships between environmental features – i.e., topological maps? In a sense, the role of the hippocampus is to associate internal and external coordinate systems and to accommodate cue conflict situations (re-instantiate the context when there is a mismatch between internal and external relationships). Rather than debating whether the hippocampus is the depositary of the declarative memory or the index access to a collection of maps, it may be more productive to ask just what is the role of the hippocampus in navigation and memory. With this in mind, in the article "The hippocampal debate: Are we asking the right questions?" [423], Redish suggests that there are multiple memory systems in the brain and multiple navigation systems.

8.3.1 The Cognitive Map Hypothesis

The concept of cognitive or mental map arises naturally from the place cells. The existence of these cells seem to somehow validate the Kantian conception of space as an a priori intuition independent of the experience. The cognitive or mental map can be seen as the biological realisation of the mental space that in Kant epistemology is an a priori intuition.

The concept of mental map must be seen as the tool that allows animals to explore the environment e.g., search for food or find their way back to their nests. An extreme case is to find the nest after long migrations, like for example in the Artic Tern, seabird that makes a round trip of 70,900 km between the Artic region and the Antarctic every year. It might be mentioned that the ability to build cognitive maps that represent the world was known before the discovery of place cells. In this respect, E.C. Tolman, working with rats, proved that this animal constructs field maps of the environment [424]. More precisely, Tolman demonstrated that rats have place learning ability in more or less complex mazes.

In the formation of the cognitive maps, the landmarks or external cues, play a major role. However, it might be remarked that place cells do not merely reflect sensory stimuli, like for example visual cues. It is known that for a rat in a familiar environment, the landmarks can be removed and the place cells in the hippocampus of the rat will continue firing in the same firing fields. It is evident that for an animal, to navigate in a efficient way obtaining food, hiding from potential enemies or finding the nest, it is critical to have available a mental representation of individual places. These places must be included into a metric system that allows the animal, to for example, calculate the distances between two points or know the direction taken.

The cognitive map was challenged by a more complex conceptualisation of the hippocampus as the locus of a path integrator. Information like velocity or direction is integrated into a navigational position. The omnidirectional property which is at the root of the idea of the allocentric map is not valid when the rat is placed in a unidirectional space – e.g., moving in a straight alley. In this case, place cells are not omnidirectional but direction-dependent, put simply, different neurons are active for the same place, depending on the direction taken. This finding does not discredit the idea of cognitive map, rather it forces us to question about how the maps are effectively created in the brain.

The map can not be created exclusively based on visual cues, other inputs must be considered, like for example speed or direction. The cognitive map mechanism needs to be thought in a wider context, to be incorporated into a more dynamic and systemic perspective. A map is a static object that does not match in a changing world. Besides, there is the problem of the overlapping maps stored in the very same population of place cells. If we think the problem of navigation and spatial representation in terms of dead reckoning or path integration, the problems take a different form. The animal needs are other than visual cues or landmarks, self-motion is necessary too in order to have a metric that allow the animal to calculate distances and optimal routes.

8.4 The Hippocampus: Anatomy and Connectivity

In order to understand the function, one has to deal with the structure first. This motto is elaborated here and both the anatomy and the architecture of the hippocampus are explored.

The hippocampus is an allocortical structure in the medial temporal lobe. The most studied cell in the nervous system is the pyramidal cell of the hippocampus. One of the main reasons why it has been invested so much time and effort in investigating the anatomical and functional aspects aspects of the hippocampus is its simple (if we compare it with the six-layered cortical structure) architecture of three layers. The modern name to refer to the allocortical structures of the brain is limbic system. The limbic system includes the amygdala, the hippocampus, the entorhinal cortex and the hypothalamus. It is commonly assumed that the limbic system plays a major role in the emotional aspects of the animal, so emotions and feelings like love, hate, fear ... would be "located" in the limbic allocortical areas of the brain. It is more correct to talk about hippocampal formation than hippocampus. The hippocampal formation is thus, a group of brain areas consisting of dentate gyrus, hippocampus, subiculum, presubiculum, parasubiculum and entorhinal cortex.

The pioneer work of David Marr in computational hippocampal models, assumed the unidirectionality of connections. Anatomists today have shown that this assumption is not completely accurate, there is bidirectionality between, for example, the dentate gyrus (DG) and the Cornu Ammonis region 3 (CA3). However, the flow of information is largely unidirectional. The general picture of the hippocampus connectivity is as follows: the signal input to the hippocampus from the cortex arrives via the entorhinal cortex (EC) and other parahippocampal regions. The output projects back to parahippocampal cortices, primarily via the subiculum and projects to subcortical structures via the fornix. The output signal is not conveyed exclusively to the EC, additional output pathways go to other cortical areas like the prefrontal cortex, and large output goes to the limbic system, for example, the lateral septal area. An original drawing of made by Cajal is shown in Fig. 8.1.

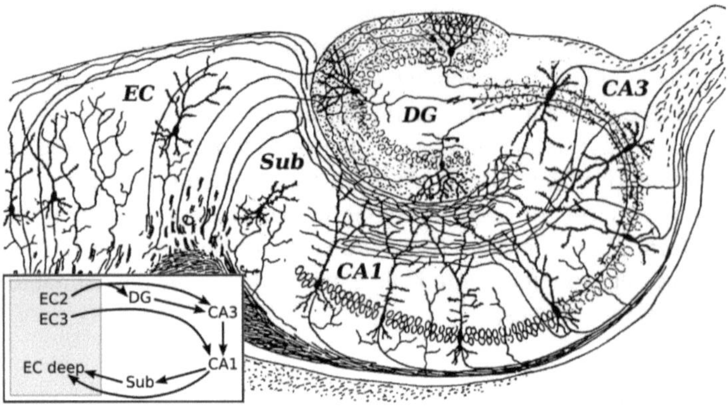

Fig. 8.1 The hippocampus and its circuitry drawn by Ramón y Cajal. The superficial layers of the EC (layers 1,2,3) provide the most prominent input to the hippocampus, and the deep layers (layers 4,5,6) of the EC, receive the most prominent output

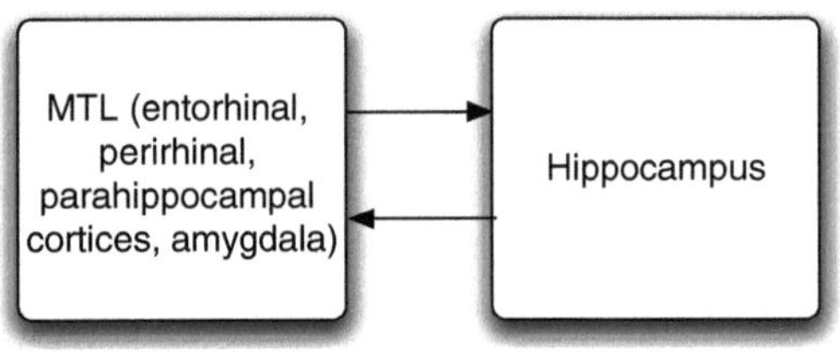

Fig. 8.2 Simplification of the usual schema of connections modules in the hippocampus. The medial temporal lobe (MTL) is influenced by and influences the hippocampus proper

This scheme of connections depicted in Fig. 8.1 can be simplified if we cluster the regions: entorhinal, perirhinal, parahippocampal cortices and the amygdala under the label medial temporal lobe (MTL). The MTL regions shape hippocampal processing and are in turn often shaped by it (Fig. 8.2).

8.5 Place Cells

Place cells are neurons located in the brain hippocampus that fire in complex bursts whenever a rat, moves through a specific location in specific location. The striking thing about place cells is that they code the spatial position of the animal, irrespective of either the direction from which the position is reached or the behavior of the rat at any precise instant. Thus, there is a direct link between the neural activity of a single cell and a Cartesian position of the rat. How does the animal know that it is in a particular position? Apparently this could be done by computing the allocentric space, landmark or visual cues. The most important property of these place cells is their omnidirectional property, that can be observed in the conical

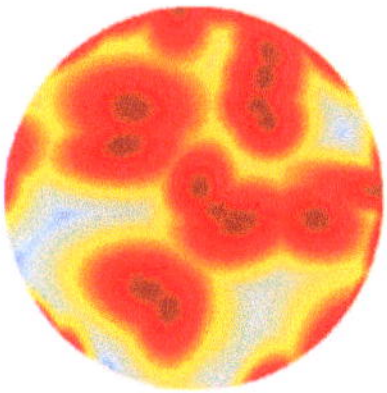

Fig. 8.3 The picture shows the 36 place fields one for each of the 36 rat hippocampal CA1 place cells. Thus, as it is shown in the figure the position of the rat is contained in the firing of the place cells. The place fields are cone shapes, this means that the firing rates increases irrespective of the direction from which the rat arrives

shape of their activation landscapes (the firing rate increases when the rat approaches the location, independently of the direction is heading when it does it). Thus, the immediate conclusion is that place cells are coding explicit (no contextual) locations in the environment and not particular sensorial cues.

The region in which a place cell fires the most is called its place field. Thus, there is a correspondence place field/place cell. What defines a place field is that the firing rate within the field is much higher than outside (it can be as much as 100 Hz. from virtually 0). For a given environment, we can determine a collection of place cells whose associated place fields cover the whole environment. There is enough evidence that under certain circumstances, the hippocampal pyramidal cells show location associated firing. However, it is less clear what they really represent; there are those who argue that place cells can be an epiphenomenon, produced by the spatial nature of the experiments where these cells are discovered. Granted that place cells are correlated to space, the question that arises is: are the place cells the only neurons correlated to space? The possible representational content of these cells and of the assemblies they constitute, can serve to further question how the hippocampus contributes to spatial representation, navigation and episodic memory.

8.5.1 Place Cells as Representational Entities

The interest in these cells is rooted in the fact that they are candidates to be the direct representation of the external space. A place cell, fires maximally when the animal is in a specific position or place field, so the firing rate of a cell can be used to decode the position of the animal within the environment with striking accuracy (Fig. 8.3). The existence of place cells was not accepted until Muller et al. [425] came out with the numerical method that allow to quantify the place fields. At this point we need to formally define the term place field: a place field for a place cell is a space $\mathcal{F} \in \mathbb{R}^2$ with significative firing rate. Thus, a place field $\mathcal{F} \in \mathbb{R}^2$ is such that $fr(\mathcal{F}) > k$. Being k a constant that represents the threshold or firing rate for part of the space that are no place fields and fr a function that returns the minimum firing rate for all the pixels of the space $\mathcal{F}$. The environment is totally covered, which means that for every pixel p of the entire environment $E \in \mathbb{R}^2$, it holds $p \in \mathcal{F} \cup \bar{\mathcal{F}}$

8.6 Grid Cells

Grid cells, likewise place cells, are place-modulated neurons. The firing location of a grid cell is multiple, contrary to the place cells which are mono field. The multiple firing location of a grid cell is indeed a grid with a most striking property, it is an array of equilateral triangles (Fig. 8.4).

It might be noted that grid cells were discovered, while researchers investigated whether place cells activity was endogenous to the hippocampus. The hypothesis was that CA3 and DG are the intra-hippocampal inputs to CA1, which is the area in the hippocampus where one can find most of the place fields. This idea was proven wrong, after removing CA3 and DG, that is, the main CA1 inputs, the spatial firing in CA1 persisted. So place signals did not exclusively arise within the hippocampus, spatial signal was brought to the CA1 from outside the hippocampus.

In 2004 Fyhn et al. [426] discovered a group of neurons in the medial entorhinal cortex (mEC) that shows spatial related firing. The mEC cells have a sharply tuned spatial firing, much like the hippocampal place cells do, but with one difference, each of the mEC cells has multiple firing fields rather than one as is the case in the place cells. One year later, Hafting et al. [427] discovered that the several firing fields of each neuron, generate a grid. Thus, as the animal moves, the grid cells tile the environment with periodic triangles that reflect the changing position.

The grid cells have been found in the mEC, within the six layers of this cortical structure, and it is in layer II where we find the highest density of this kind of cells. The neurons in the layer II of the medial entorhinal cortex (mEC-II) are the main input of the place cells, but in contrast to the place cells, the entorhinal cells are activated throughout the environmental terrain, whenever the animal is at the vertex of some equilateral triangle, forming a tessellation or grid. In short, both place cells and grid cells are neurons with spatial located firing, in other words, they have spatial representational power, allowing the animal to know its position and to navigate in an environment, for example find the way back home after eating. The difference, apart from the fact that place cells are hippocampal neurons and grid

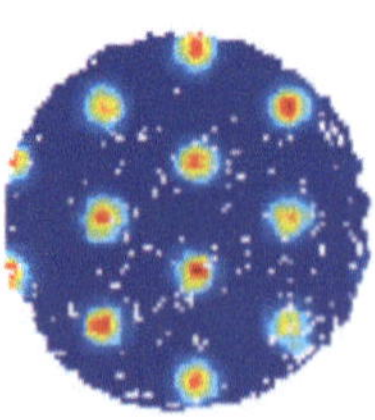

Fig. 8.4 Grid map of a rat running on a linear track after 10 min [23]

Table 8.1 Place cells and grid cells, similarities and differences

	Brain area	Type of map	Activation
Place cells	Hippocampus	Static	Need input from mEC
Grid cells	mEC	Dynamic	Active instantaneously in any novel environment

cells are in the mEC, is that whereas a place cell has a single firing location, a grid cell has multiple firing fields with a striking geometric regularity; the firing fields form periodic triangular arrays, tiling the entire environment available to the animal.

8.6.1 Grid Fields

Three parameters totally describe the grid of a cell (grid cell): *spacing* is the distance between contiguous fields, *orientation* the rotation angle of the grid referred to the reference axis, *spatial phase* is how much the grid is translated relative to an external reference point.

A grid field for a grid cell is a set of open balls $G_i : 1..n$, where for every ball G_i, $fr(G_i) > k$, that is to say, it has a significative firing rate. Thus, so far, the definition of G_i is identical to place field seen in Sect. 8.5.1. Additionally, every ball G_j of a grid field, form a equilateral triangle with its two other closest balls G_i and G_k in $\cup_1^n G_i$ or G to abbreviate. The grid field G is identified by the three parameters, spacing, orientation and phase that can be trivially obtained from the metric of the space defined above.

The processing of the place signal is therefore not an exclusive privilege of the hippocampus, the mEC plays a major role in the spatial representational tasks. The majority of cells in mEC-II and mEC-III have grid properties, this means that most of the cortical input to the hippocampal place cells that have to do with spatial representation that come from grid cells in the mEC. Grid cells can be found just one synapse upstream of the place cells [427]. Therefore, acknowledging that grid cells and place cells are intrinsically connected, to claim that place field may be extracted from grid fields [428] deserves to be appropriately explored (Table 8.1).

Mathematically, using Fourier's analysis, several grid fields with different spacing can combine linearly to yield a place field. Solstad et al. [429] proposes a computational model showing that a place field can arise by the sum of 10–50 grid cells. When the spatial phase variation in the grid-cell input was higher, multiple and irregularly spaced firing fields were formed. This idea has been very appealing in the hippocampus community, and it has helped to produce a large number of models with a common tenet: place cells in the hippocampus compete to receive the summed firing pattern activity of the cortical grid cells. The problem with these kind of models that transform grid patterns into place patterns is that they do not tell us that much about the mechanisms that underlie the spatial firing pattern of grid cells and place cells. Besides, it is debatable that a linear sum of grid cell patterns underlying a specific metric is the correct way to model a place cell pattern which represents topologically the environment without metrical relations. It might be remarked here that the models of grid field formation, deal with timing rather than with structure or connectivity, and this is because they assume that the structure is already known. In these models the structure is a single cell whose firing activity needs to be understood. There is a number of computational models that aim to simulate a grid field, however they do not tell us much about the causes that originates that phenomenon, let alone a mechanistic explanation of the emergence of place cells in the hippocampus. The study of grid cells is still in its infancy and the mechanisms that underlie the spatial firing pattern are still waiting to be discovered [430].

8.7 Head Direction Cells

Head direction cells are neurons with direction-based firing, that is to say, these cells fire when the head of the rat is pointing to a particular direction regardless of the location of the rat or even the angle of the neck. The head direction neurons were initially found in the postsubiculum (between the hippocampus and the EC) [431] and they have been found in a number of areas, including limbic structures [432, 433] and the striatum [434]. It has been conjectured that head direction cells work as a compass, but the attractor magnetic field is not the earth's magnetic pole but some arbitrary reference direction. They provide information of the direction that may be integrated with information of the speed and location and would allow the animal to navigate using path integration.

The direction system poses important problems to the cognitive map theory. Experiments with rats in rectangular boxes showed that place cells do not remain invariant to the modification of the box. Indeed place fields are malleable, they change sizes, split into two and even disappear. O'Keefe suggestion to this disconformity with place cell location selective discharges that himself discovered is that the rat calculates the size of the box by triangulation on the basis of visual cues e.g., the heights of the wall [435]. McNaughton et al. [436] less confident on rats geometrical skills, suggest that distance is calculated based on self-motion cues. As the rat moves, it remembers the physical contact with the walls and compute the vectorial distance (amount and direction of movement). The idea is that direction is obtained outside the hippocampus and delivered to the place cell system. The hippocampus would act as an integrator that calculates shortest distances between landmarks based on direction, velocity and position information. The hippocampus navigational system must take into account other cues, notably the velocity, for [437] the firing rate of single cells is a function of location and also speed (Fig. 8.5).

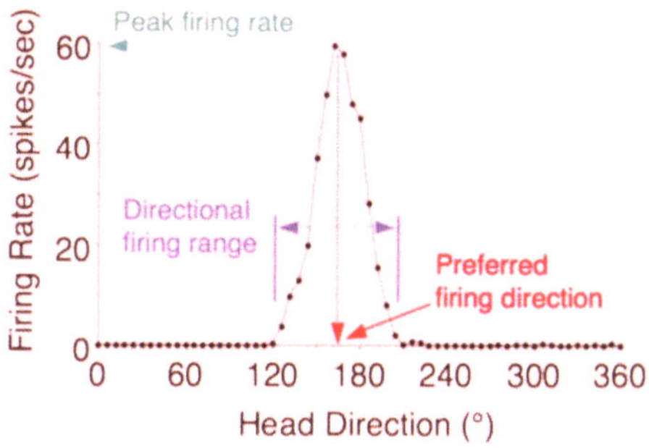

Fig. 8.5 Firing rate vs. head direction plot that shows the preference of this cell around the 180° (After [24])

8.8 A Theory of Brain Spatial Representation Based on Category Theory

The huge amount of information on brain structure and operation, in order to have lasting impact in brain theory, must be analysed within consistent theories. Otherwise the global picture of brain operation will be missing. It is necessary to look for the fundamental structures which must be found not just in the objects – the neurons – but also in the relationships between the objects and the emerging organisations. How components at the same level interact (the objects here being neurons) and how superior levels constrain those levels below and emerge from those above is tackled here with a mathematical tooling. The mathematical theory of categories is proposed as a valid foundational framework for theoretical modeling in brain sciences.

8.8.1 The Category of Neurons

One of the highlights of this work is that it exemplifies the theory of categories in strong non algebraic categories. Indeed, the crucial aspect and novelty in this approach needs to be met in the categorical construction of biological (non algebraic) categories. For this purpose we must find a definition for a neural abstract category **CAT-Neur** as a category whose objects are either neurons or sets of neurons. **CAT-Neur** as any other category, consists of three things, (i) a set of objects O, (ii) a set of morphisms $Mor(A, B)$[2] for any two objects A,B of O, and (iii) a rule of composition that fulfills the properties of associativity and identity.

[2]We use $Mor(A, B)$ rather than he notation used in Chap. 6, $Hom_{CAT-Neur}(A, B)$, because is shorter

We identify three possible categories for the category **CAT-Neur** that are useful for the development of the theory introduced in this chapter.

(i) The category **Neur**.
(ii) The category **Neur+**.
(iii) The category **Neur***.

The category **Neur**, whose objects are neurons and the morphisms are the synaptic paths between them, with the convolution of paths as composition.

The category **Neur*** which is the category of neurons where the objects are topological spaces of neurons (N, θ) and the morphisms are continuous maps. A function from two topological spaces $f : (N, \theta) \to (M, \upsilon)$ is continuous if $f^{-1}(B) \in \theta$ whenever $B \in \upsilon$.

The category **Neur+**, which has as objects metric spaces and as morphisms Lipschitz maps for $\lambda = 1$ that preserve distances. Note that a Lipschitz map is always continuous but the contrary is not true. The morphisms in **Neur+** preserve distances between metric spaces which exemplify neural assemblies.

8.8.2 The Category of Places

Now we will define a category for modeling place fields, that is the physical locations that produce the spike firing in the grid cells and place cells. Following the previous definition for **Neur**, the category **Field+** has as objects metric spaces (including hexagonal grids) and as morphisms contractions (a specific class of functions between metric spaces), and the category **Field*** is composed of topological spaces as objects and continuous functions as morphisms.

The category of metric spaces is of course defined by objects and morphisms. The objects are metric spaces (X, d) and the morphisms are mappings between metric spaces $(X, d) \to (X', d')$ (Fig. 8.6). As in any other category, the composition of morphisms must satisfy associativity and identity. Let us define first the objects i.e., metric space. A metric space is a structure (X, d) with X being a set and the function $d : X \times X \to \mathbb{R}^+$ satisfying:

1. $d(x, y) = 0$ when $x = y$
2. $d(x, y) = d(y, x)$ and
3. $d(x, z) \leq d(x, y) + d(y, z)$

The Euclidean distance is a map $d : \mathbb{R}^n \times \mathbb{R}^n \to \mathbb{R}^+$. For $n = 2$ the distance is $d((x_1, y_1), (x_2, y_2)) = \sqrt{((x_1 - x_2)^2 + (y_1 - y_2)^2)}$. Typically the function d is assumed to be the Euclidean distance, but of course, other distances are possible. One example of a metric that satisfies the three axions above is the "Manhattan distance" or $d : \mathbb{R}^n \times \mathbb{R}^n \to \mathbb{R}^+$, so for a two dimension space, $d((x_1, y_1), (x_2, y_2)) = |x_1 - x_2| + |y_1 - y_2|$.

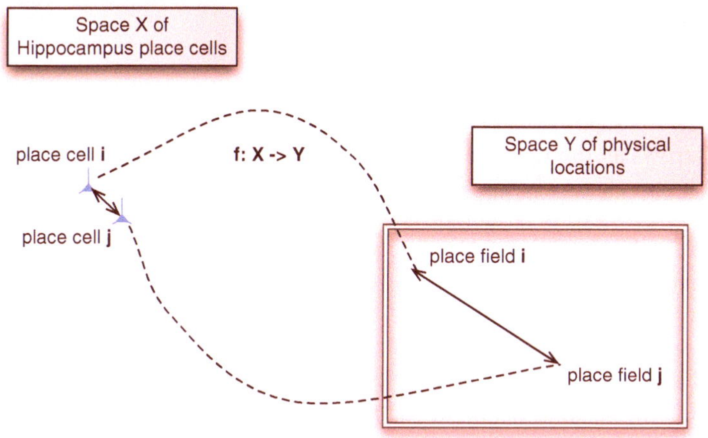

Fig. 8.6 (X, d) is a metric space where X is the set of place cells in the hippocampus and d the Euclidean metric distance, (Y, e) is a metric space in the bidimensional plane with identical distance $e = d$. The mapping between the metric spaces $f : X \to Y$ preserves the distances if $e(f(x_1), f(x_2)) = d(x_1, x_2)$. f is said to be an isometry and is immediately a monomorphism (Demonstration: $x_1 \neq x_2, e(f(x_1), f(x_2)) = d(x_1, x_2) \neq 0 \Rightarrow f(x_1) \neq f(x_2)$). Thus, an isometry that is an epimorphism is an isomorphism. Note that empirically speaking, it makes no sense to take the space (Y, e) as a metric space, for example, closed points in the space are not necessarily mapped onto closed place cells in the hippocampus. Thus, objects in the category of neurons can not be mapped directly onto metric spaces

Definition 8.1. A mapping $f : (X, d) \to (X', d')$ preserves distances if for all pair of points, $x_1, x_2 \in X$, it holds $d(x_1, x_2) = e(f(x_1)f(x_2))$.

Definition 8.2. A function $f : (X, d) \to (Y, e)$ between two metric spaces is continuous at $x_0 \in X$ if for all $\epsilon > 0$ there exists $\delta > 0$ such that if $d(x, x_0) < \delta$ then $e(f(x_0), f(x)) < \epsilon$

A contraction is a Lipschitz map with $\lambda < 1$, while a map between two metric spaces $f : (X, d) \to (X', e)$, is such that $d(x_1, x_2) = e(f(x_1)f(x_2))$, is a distance preserving map . Note that every Liptschitz function is continuous and as a contraction is a Lipschitz map with $\lambda < 1$, contractions are therefore continuous. For the demonstration that Lipschitz functions are continuous see [173].

Now we are able to define the category **Met** of metric spaces and Lipschitz maps that are structure preserving maps. The composition of Lipschitz maps, gf, is a Lipschitz map and the properties associativity of composition and identity $id_x : (X, d) \to (X, d)$, are trivially demonstrated. The topological spaces are useful when we are interested in closeness and continuity rather than in distance as it is the case in metric spaces. The category of topological spaces **Top** is one that has topological spaces as objects and continuous maps as morphisms.

8.8.3 Functor Between Neur and Field

At this point we wish to define the functor between the categories that have been defined. Let us suppose that **Neur+** is a category whose objects are sets of neurons and the arrows all the functions between them, so in the case that there is only one place cell, the category **Neur+** is a set of a single element or singleton. For an object of category **C**, there is an unique functor $F : \mathbf{C} \to \mathbf{1}$. Thus, there is an unique functor from the category of metric spaces and Lipschitz-distance preserving maps, **Field+**, and the category of one place cell **1**.

Functors preserve isomorphisms, so given the functor $F : \mathbf{C} \to \mathbf{D}$, the isomorphisms in category **C** are preserved in category **D**. An interesting feature of functors is that they may preserve properties. For example, since functors preserve composition of morphisms $\circ$, and identities, id, they preserve every property that can be positively expressed in the language of $\circ$ and id. In particular, functors preserve commutativity of diagrams [438]. So given a functor $F : \mathbf{C} \to \mathbf{D}$, for certain objects, arrows or composition of arrows in category **C**, that have the property p, the functor F brings such property to the F-image.

Definition 8.3. Let **C** and **C**$'$ be two categories, a covariant functor F from **C** to **C**$'$ is defined as a rule which associates for every object A in **C** an object $F(A)$ in the category **C**$'$ plus a rule that associates for every morphism $\alpha : A \to B$ in **C** a morphism $F(\alpha) : F(A) \to F(B)$ in category **C**$'$. Then F must satisfy the following two conditions:

1. The composition is preserved: for the diagram $A \overset{\alpha}{\to} B \overset{\beta}{\to} C$ in **C**, $F(\alpha \circ \beta) = F(\alpha) \circ F(\beta)$
2. Identities are preserved: for any object A in the category **C**, $F(id_A) = id(F_A)$

Now, the functor or more precisely a covariant functor from the category of neurons **CAT-Neur** to the category of metric spaces **Met**, F : **CAT-Neur** $\to$ **Met** is such that (i) and (ii) hold.

(i) Every object N in the category of neurons **CAT-Neur** is mapped onto an object $F(N)$ in the category **Met**,

(ii) Every morphism $\alpha : N \to N'$ in **CAT-Neur** is mapped onto a morphism $F(\alpha) : F(N) \to F(N')$ in the category **Met** and F preserves composition and identity,

(ii.a) The composition is preserved, so $A \overset{\alpha}{\to} B \overset{\beta}{\to} C$ in **CAT-Neur**, $F(\alpha \circ_N \beta) = F(\alpha) \circ_M F(\beta)$ (both sides of the equation are morphisms in **Met**)

(ii.b) Identities are preserved, so for any object A in the category **CAT-Neur**, $F(id_A) = id(F_a)$ (both sides of the equation are morphisms in **Met**)

The physiological interpretation of functor F is as follows: (i) means that it is possible for any object N in the category of neurons **CAT-Neur** to have associated a metric space (X, d). Note that this is different to assign a location to a set of neurons, rather we are associating a set of neurons with a metric space $N \rightarrow (X, d)$, where $d : X \times X \rightarrow \mathbb{R}^+$. For example, let **Met1** be the category of metric planar space of diameter 1, (M, υ), that is, $d(m, m') \leq 1$ for all $m, m' \in \mathcal{M}$, being $\mathcal{M}$ an open ball. Then $F(N)$, $F : N \rightarrow (M, \upsilon)$, represents that the object N, a set of neurons, falls into a circumference of diameter one in the two-dimensional space M. On the other hand, if we take for the category **CAT-Neur** the category **Neur**, then condition (ii) can be interpreted as follows: whenever there is a synapse between two neurons n, n', $\alpha : n \rightarrow n'$, there is a relationship between the metric spaces associated to each of the synaptic neurons, $F(\alpha) : F(N) \rightarrow F(N')$, such that F is a map that preserves composition and identity.

Let $A \overset{\alpha}{\rightarrow} B \overset{\beta}{\rightarrow} C$, then $F(\alpha \circ \beta) = F(\alpha) \circ F(\beta)$ simply means that the map associated to a synaptic path is equivalent to the map associated to the synapses. The other requirement, identity is preserved, can be interpreted as there is always a metric space for any neuron.

It might be remarked that the functor F defined here, does not preserve the metric space defined in the category **Met**. This is in accordance with the empirical fact that the brain has no metric or at least not a Euclidean-like metric based on the distance. Indeed, what F does is to bring the structure of the category of neurons over to the category of metric spaces **Met**. The very different nature of the two categories that are being mapped by F, makes difficult to see how F works. With an example we will try to make this point more clear. Let the objects of **Neur** be place cells, that is, neurons that fire when the brain occupies a position in a plane surface, for example, inside a maze or a box. The metric space for the environment is given by the category of metric spaces **Met**. For every synapse α coupling two place cells, N and N' in **Neur**, $F(N)$ and $F(N')$ represent the place fields of N and N' respectively in **Met** Thus, the mapping F, in order to be a functor needs to be a structure preserving map between the two categories being mapped by the functor i.e., **Neur** and **Met**. In the case that **CAT-Neur** is **Neur** whose objects are neurons, the relationship between the place field of the postsynaptic cell $F(N')$ and the place field of the presynaptic cell $F(N)$ may be exemplified by $d(F(N_i), F(N_j)) \leq d(N_i', N_j')$, where N_i, N_j are in category **Neur**, and N_i', N_j' in category **Met**.

8.9 A New Framework for Place and Grid Cells

Here we propose a radically new theoretical framework for the formation of place cells from grid cells. The computational models of the hippocampus [439–441] state that the sum of a set of elements, grid cells, directly produce another element, a place cell. In doing so, these models take for granted that the properties of the sum are directly reducible to those of its components. This strict form of reductionism is at odds with the nature of complex systems. It is necessary to tackle the system as a whole, and bring to light the way in which the components interact, producing higher levels of complexity, exemplified in new systemic properties that are not present in the single components. It might be remarked here, that this is not a criticism of the reductionist approach. Indeed the reductionist analysis is arguably the best plan of action that one may follow in order to understand how a system works. But this is just the half of the work, the synthetic endeavor must follow after the analysis.

In what follows, it is described the effect of injecting the concepts of co-product and colimit from category theory into the problem of place cell formation in the hippocampus. The classical reductionism credo states that the whole is no more than the sum of its parts. Therefore the properties of the sum are reduced to those of its components, without introducing new properties. This is what the categorical concept coproduct exemplifies.

Definition 8.4. A coproduct of two objects A and B is a an object $A + B$ together with the arrows $A \xrightarrow{\iota_1} A + B$ and $B \xrightarrow{\iota_2} A + B$, such that for any object C and the pair of arrows $A \xrightarrow{\alpha} C$, $B \xrightarrow{\beta} C$, it exists an unique morphism π that makes the following diagram commutes.

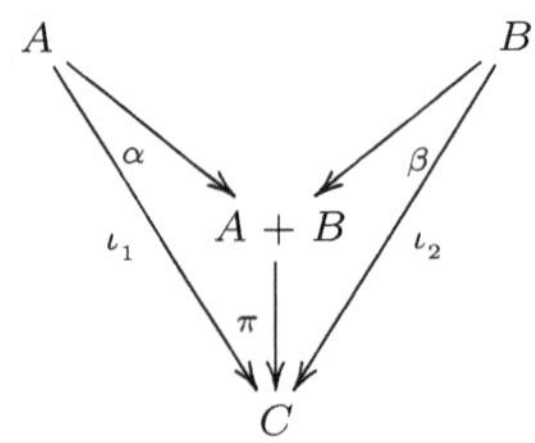

A generalised diagram of coproduct also called direct sum is shown next.

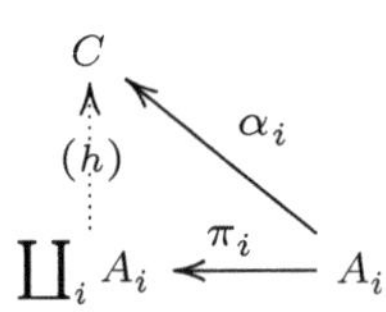

Thus, all one needs to know is the components A_i that participate in the coproduct $\coprod_i A_i$. This is possible because all the components play a symmetrical role in the construction coproduct.

On the other hand, the colimit cP embodies the collective operations made by the family of components A_i which are made possible because the components cooperate by means of the links that connect them [184]. The colimit in a category of a family of components A_i without any arrow between them is the coproduct. The colimit, contrary to the coproduct, entails a non symmetric relationship with its components. As Fig. 8.7 depicts, the coproduct can be compared to the colimit cP. This symmetry breaking process may be somehow quantified by the arrow s.

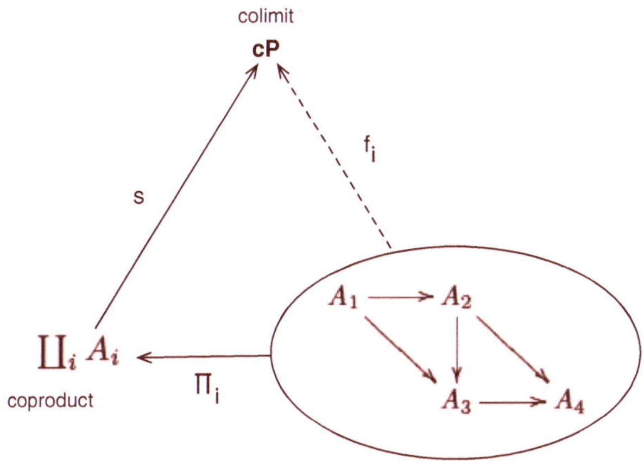

Fig. 8.7 The family of objects A_1, A_2, A_3, A_4 has both a colimit **cP** and a coproduct $\coprod_i A_i$. The coproduct is linked by s to the colimit. The link s express the transit from the coproduct to the colimit and embodies the symmetry breaking in the relationship between the family of objects A_i and the colimit

Fig. 8.8 A colimit cD for the base diagram D with three objects A_1, A_2, A_3

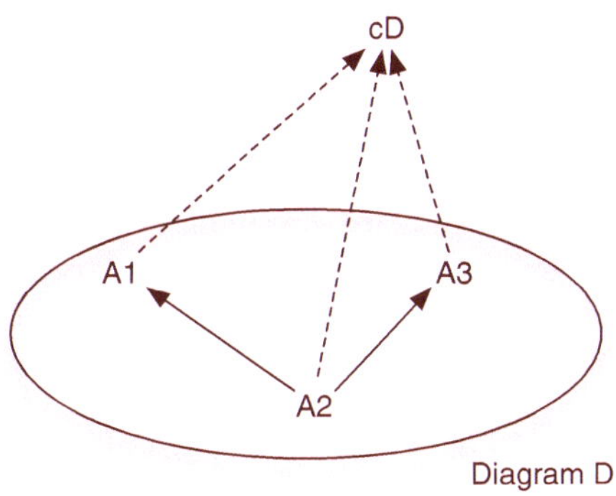

8.9.1 *Place Field as Colimit of Grid Fields*

The hypothesis posed here is that the cooperation of several grid cells identified with grid fields gives rise to the colimit which is a place field. Thus, the colimit of the metric system depicted in Fig. 8.8 can be seen as an integrator of the information contained in the metric system components. It might be remarked that the colimit is an object of the category Field, a sort of complex object that actualizes the internal organisation of the objects that it is binding. Colimits and limits do not exist for all diagrams in all categories, but if they exist, they are unique up to isomorphism. The mathematical definition of colimit needs two prior definitions, namely, diagram, which was defined in Definition 3.11 and cocone. It is worth reminding here that a diagram D in a category $\mathbf{C}$ is a collection of vertices and directed edges consistently labeled with objects and arrows of the category $\mathbf{C}$.

Definition 8.5. Let D be a diagram in a category $\mathbf{C}$ with objects labeled D_i and morphisms labeled $f_k : D_i \to D_j$. We call cocone K for diagram D to the apical object B, together with the set of morphisms $g_i : D_i \to B$ forming a commutative diagram, that is, $g_j \circ f_k = g_i$

Remark 8.1. Given two cocones K' and K'' for diagram D, a cocone morphism $h : B' \to B''$ is a morphism in $\mathbf{C}$ such that $g_i^{''} = h \circ g_i^{'}$. To simplify the notation we denote the cocone morphism determined by h as $h : K \to K'$. Directly, the cocones form a category, the category of cocones $\mathbf{Coc_D}$.

Definition 8.6. A colimit for the diagram D is an initial object K in the category $\mathbf{Coc_D}$, that is, for any other cocone K' for diagram D, there exists a unique cocone morphism $h : K \to K'$.

It follows from the definition that all colimits are isomorphic because all initial objects are isomorphic.

Figure 8.9 shows that grid fields and grid cells in the medial entorhinal cortex (mEC), are linked by a map, as there is a map between place cells and place fields. Therefore for each grid cell, there is grid field, which is a metric space, where a particular arrangement generates regular hexagons. For each place cell there is one place field which is also an object of the category of metric spaces, **Field**, but in this case, its geometry is a simple point rather than a hexagon.

We can assume that the neurons – place cells and grid cells – depicted in the bottom of the figure, are in the category **Neur** having as objects neurons and as morphisms synaptic connections. However, this is not always the case. For example, a neural category whose objects contain several neurons connected between them forming populations of neurons, has neuronal assemblies as objects rather than single neurons. In this line, it is particularly valuable to shed light on how populations of grid cells contribute to the formation of one place cell. The colimit is the mathematical structure that allow us to encode the emergence of place field and the relationship between grid fields.

Now let us focus on the grid fields depicted as hexagons in Fig. 8.9 and their morphisms. It has been said above that regular hexagons are objects in the category **Field**, now we need to investigate the morphisms between the grid-field objects in this category. A contraction between two grid-field objects (G_1, d, o, ψ) and (G_2, d, o, ψ) is a continuous function $f : (G_1, d, o, \psi) \to (G_2, d, o, \psi)$, satisfying $d(f(x, y)) \leq d(x, y)$ and $o(f(x, y)) \leq o(x, y)$. This restriction is in accordance with the experimental finding that shows that spacing in grid fields increases along the dorsoventral axis in the medial entorhinal cortex (mEC). This fact appears to be correlated with the increase in size of place fields along the dorsoventral axis of the hippocampus [442, 443].

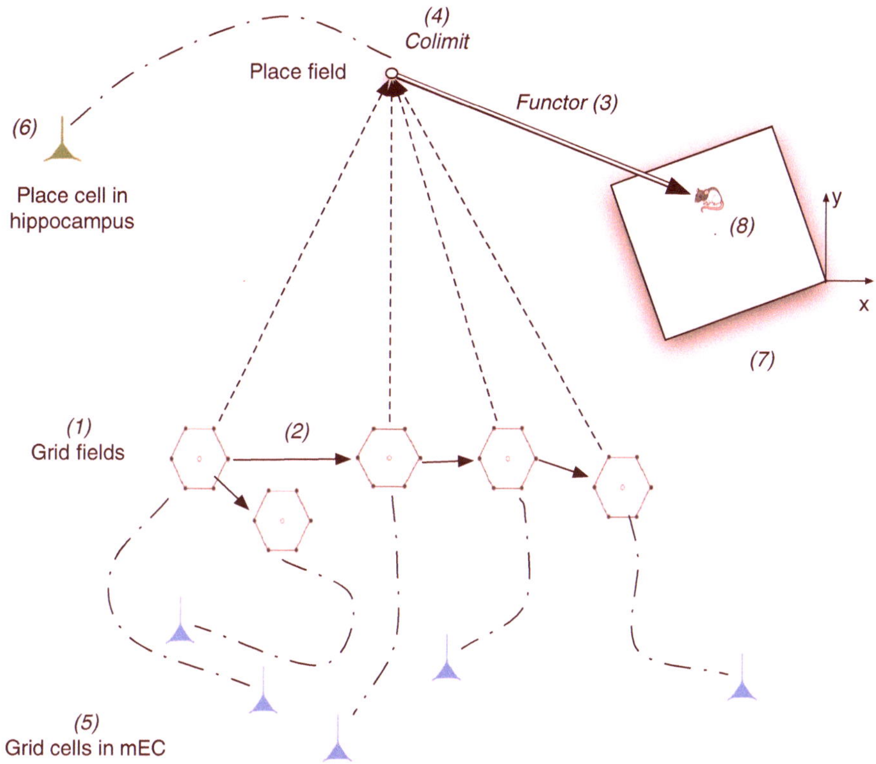

Fig. 8.9 The figure depicts a colimit where (*4*) acts as the place field of a place cell (*6*) in the hippocampus. The colimit is produced by several grid fields (one grid field (*1*) is produced by one grid cell (*5*)). (*3*) is a Functor between the colimit (*4*) and the physical location (*8*). (*2*) are connections among grid fields (*1*), which produces the tessellation of the arena (*7*)

Neighbor cells in the mEC have similar spacing and orientation. However, there is no evidence that anatomical cell clusters, correspond to functionally segregated grid maps with their own spacing and orientation [428]. On the other hand, the phase of the grid (ψ) does not follow the restriction of continuity that spacing and orientation have. Indeed, firing vertices of colocalized grid cells are shifted randomly, that is to say, the mapping between vertices in the grid field and the external reference grid is not continuous. This is in fact how fields of neighboring hippocampal place cells behave.

The colimit is a universal property which is a remarkable fact that deserves to be explained. When a mathematical construction, in our case a colimit, satisfies a universal property, one can forget the details of the structure and focus on the universal property, because all that has to be known about the structure e.g., a diagram is exemplified in the universal property. One important point that needs emphasis is that the existence of a colimit imposes constraints, not only on the diagram of grid cells that determine the colimit, but also on all the objects of the category. Besides, the colimit, if it exists, is uniquely determined (up to isomorphism) but the reverse is not true, one colimit can have several decompositions. Put it in the context of Fig. 8.9, this means that when the coordinated activity of a group of grid cells produce a place cell, this is a colimit and it is unique. But given a place cell, its place field cannot be uniquely determined by a group of grid cells, as a matter of fact, several groups of grid fields are possible for that place field.

Chapter 9
From Cells to Memories: A Categorical Approach

9.1 Introduction

This chapter can be seen as a continuation of the precedent one in which the relationship between pyramidal neurons acquires a categorical treatment. Here we extend this view to sketch a theory of declarative memory, building on the results described on the previous chapters. The original and innovative aspects of this thesis reside in its radically new insight into the structure, function and dynamics of the brain. A theory that fully explains and predicts the highly complex cognitive abilities like perception, memory or learning has not been produced yet. It goes without saying that to shed light on the role played by the hippocampal system in cognitive functions like memory and learning can be of extraordinary value for the future of cognitive and brain studies.

9.2 Types of Memory

Just as any other higher cognitive function, to try to give a definition of memory seems hopeless. The definition in the MIT encyclopedia of cognitive sciences [444] may seem open enough to satisfy everyone: "the term memory implies the capacity to encode, store, and retrieve information". However, it is also too unrestricted to provide a clear idea about what memory is and how it works. Certainly, memory is not a univocal term, it has several forms that depend on different brain mechanisms. So a well-founded strategy to get an understanding of how memory works, is to investigate how such cognitive process is implemented in the brain. Thex idea behind this is that the layman's view of memory, which is still commonly used, will become irrelevant once the biological mechanisms of memory have been uncovered and, if possible, described in mathematical terms.

Long term memory involves an connectivity changes in the brain, any new memory "engenders", somehow, new connectivity patters. Of course memory and

J. Gómez-Ramirez, *A New Foundation for Representation in Cognitive and Brain Science*, Springer Series in Cognitive and Neural Systems 7, DOI 10.1007/978-94-007-7738-5_9,

the processes involved such as learning can be tackled at different levels, for example at biochemical level, PKIZ is a molecule that plays a critical role in erasing memories. A great deal of what it is known today about memory in humans is due to one patient called Henry Muleson, better known by the initials H.M. H.M. suffered violent epileptic seizures and in 1953 the epilepsy was surgically treated by bilateral excision of the hippocampus. As a result, the epilepsy disappeared but the secondary effect was dramatic, H.M. completely lost his capacity to recognize people that he saw everyday, and the ability to remember the content of anything he read vanished after the surgery. Scoville and Milner [445] concluded that the hippocampus, as the H.M. case clearly showed, was related to memory functions. In essence, the hippocampus made long term memories out of short term ones. H.M. could remember his childhood and all the important events that happened prior the operation, but he could not hold on new information for more than a few minutes. The damage in the hippocampus and some surrounding structure in H.M. entailed the impossibility to create new episodic-declarative knowledge. Nevertheless the patient's procedural memory remained intact. In other words, H.M. could remember motor skills, and this explained that the manual works requested to H.M. in the laboratory improved with practice; but he could not remember at all what he did 1 hour ago. The distinction between declarative memory and procedural memory was established based on the conditions observed in H.M.

Declarative memory can be episodic or semantic, the former encode events of one's past and the last encodes semantic knowledge that can be consciously retrieved and formulated. Declarative memory is involved in modeling the external world, that is to say, in storing representations of objects, episodes, and facts. Episodic memory underlies the capacity to reference personal experiences in a given context of time and space. Semantic knowledge, on the other hand, is largely a context-free kind of memory. An example of semantic memory is building a class of similar items; if we know how a chair looks like, we will be able to, every time we see a chair, recognize it as another element of the class chair, and we do so without remember the details of the first chair we ever saw which was the first instance in the class chair. Put it simply, the semantic memory is the meaning of things.

Procedural memory is also called non declarative memory and is more an umbrella term for all those memories that are not declarative. Procedural memory includes procedural skills and the development of habits through repetition of appropriate ways to respond to stimuli. It might be noted that procedural memories are not real memories at least not in the idea of memory reminiscent to Marcel Proust *madeleine*, as the retrieval of an event from the past, rather they are skills, for example, driving a bike or playing video games i.e., skills that are done increasingly better as a result of the experience. Another major difference between declarative and procedural memory is that while declarative memory is believed to depend on the hippocampal-entorhinal system, procedural or non declarative memory is not a construct of the hippocampal system itself, rather it involves other areas like the amygdala or the cerebelum. In some sense, memory is the persistence of perception. Memory, knowledge and perception are three terms that refer to a single thing, the formation of brain systems with significant informational content that in general can be consciously retrieved.

$$Memory \begin{cases} Declarative\ memory & \begin{cases} Episodic \\ Semantic \end{cases} \\ \\ Procedural\ memory \end{cases}$$

9.3 A Theory of Declarative Memory Based on Category Theory

The dual role of the hippocampus in formation and retrieval of concepts is not surprising, especially considering that the formation of new memory (knowledge) requires the retrieval of the old one. Thus, memory is knowledge, and perception is a condition of possibility of memory and therefore of knowledge. We must exploit the important fact that from the point of view of neurobiology, memory and perception share the same neural substrate. The main point that is being explored in this section is that despite the diverse nature of episodic and semantic memory, it is possible to connect them via categorical objects like product, pullback or colimit.

Let us begin by the introduction of the categorical concept of product and its application in a navigational task in one dimension, after the results will be expanded to navigation in a two-dimensional arena and the use of the categorical concept pullback. The chapter finishes with a general formulation in categorical terms of the creation of semantic memories from episodic ones.

9.3.1 Categorical Product in Acquisition of Middle Point Concept in 1D Navigation

Suppose a rat is placed in a track (one dimensional environment), the animal immediately starts moving back and forth in order to get and idea of the dimensions of the environment. As the rat moves from one point to the other, episodic memories are created. The animal, as it starts wandering, begins to build associations of self-centered information with the temporal order in which the different positions are reached. Episodic memories are not explicit. Explicit ones may be retrievable independent of the internal state of the rat. Suppose also that there is no particular visual or smell stimulus that can make the rat remember any particular position. One may think that after a while, the rat will acquire an explicit memory, for example the concept of middle point which exemplifies the position in the track, from where it needs the same amount of time to get any of the extremes.

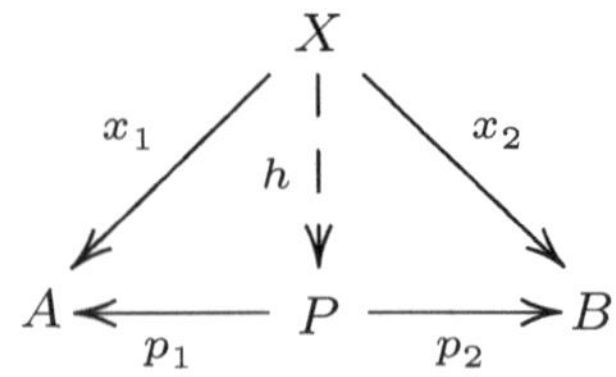

Fig. 9.1 The object P together with the *arrows* p_1 and p_2 is the product of objects A and B in a category C

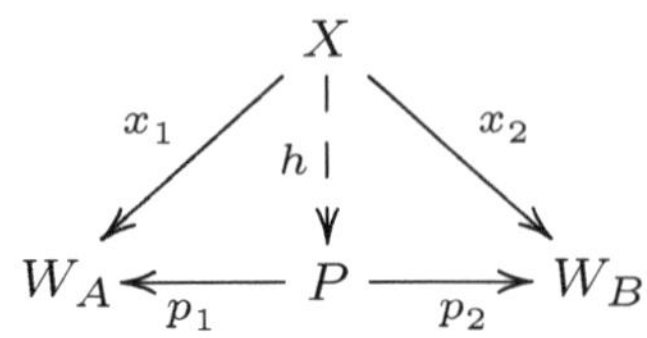

Fig. 9.2 The object P together with the *arrows* p_1 and p_2 is the product of objects W_A and W_B that represent the respective walls in the category C

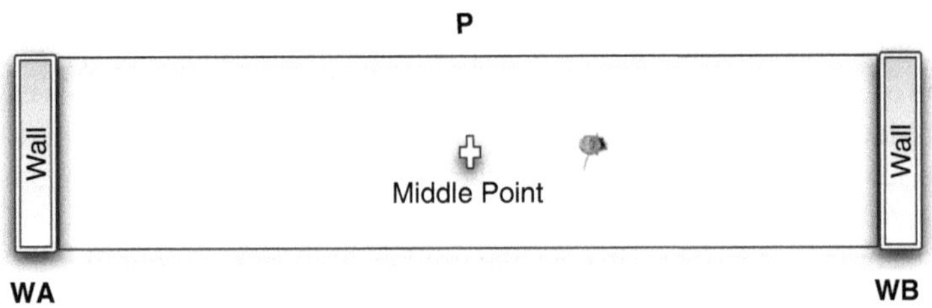

Fig. 9.3 $\mathbf{W_A}$ and $\mathbf{W_B}$ are the two walls that the rat will meet when moving in a single track maze. After reaching both walls, the animal would develop the concept of middle point **P**

Any cognitive behavior involves integration of information. The categorical concept of product is a formalisation of integration. Moreover, as it will be shown later, a product in a category that admits a final object, is an instance of a more general categorical form, namely a pullback.

Definition 9.1. In a given category **C**, a product of two objects A and B, is another object P equipped with two morphisms, $P \xrightarrow{p_1} A$ and $P \xrightarrow{q_1} B$, such that for any pair of morphisms, $X \xrightarrow{x_1} A$ and $X \xrightarrow{x_2} B$ there is an unique morphism h making the Fig. 9.1 commutes.

Note that the broken arrow h means that it is unique, and the morphisms p_1, p_2 are usually called projection morphisms. The main characteristic of a product is that the constituents are retrievable via the projection morphism. Figure 9.2 indicates the categorical product for the acquisition of the middle point between two walls W_A and W_B depicted in Fig. 9.3.

For our purpose, the categorical product given by the object P and the morphisms p_1, p_2 is a statement about a cognitive behavior of the rat, whereas X and x_1, x_2 is a constraint on what constitutes a valid product, rather than a specific claim about cognition. Note that there is not any particular commitment in the morphisms p_1, p_2.

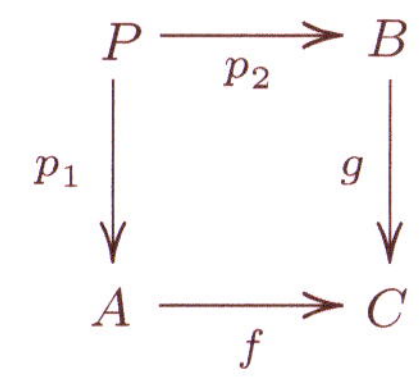

Fig. 9.4 The object P together with *arrows* p_1 and p_2 is a pullback in C

In fact, p_1 can represent the travel time to reach the wall A, W_A, but also the number of steps needed.

Figure 9.3 represents a possible experimental setting that could be use to explore the biological plausibility of our theory in the acquisition of the middle point concept in a rat moving in a single track maze. Here P, W_A and W_B are objects in the category **C** of memories or mental objects that was described in Chap. 3.

9.3.2 *Categorical Pullback in Acquisition of Middle Point Concept in 2D Navigation*

Now suppose the rat is removed from the one dimensional track depicted in Fig. 9.3 and put upon a plane. The rat's capacity to build the explicit memory for the middle point of the arena can be seen as analogous to the generalised product – i.e., a pullback.

Definition 9.2. In a category **C**, a pullback of two morphisms, f and g, with common codomain $A \xrightarrow{f} C \xleftarrow{g} B$ is an object P together with a pair of morphisms $P \xrightarrow{p_1} A$ and $P \xrightarrow{p_2} B$ that form a commutative diagram $f \circ p_1 = g \circ p_2$ (Fig. 9.4).

Moreover, the morphisms are universal among such squares because for any pair of morphisms $Z \xrightarrow{z_1} A$ and $Z \xrightarrow{z_2} B$ such that $f \circ z_1 = g \circ z_2$, there is an unique morphism h such that the following diagram commutes.

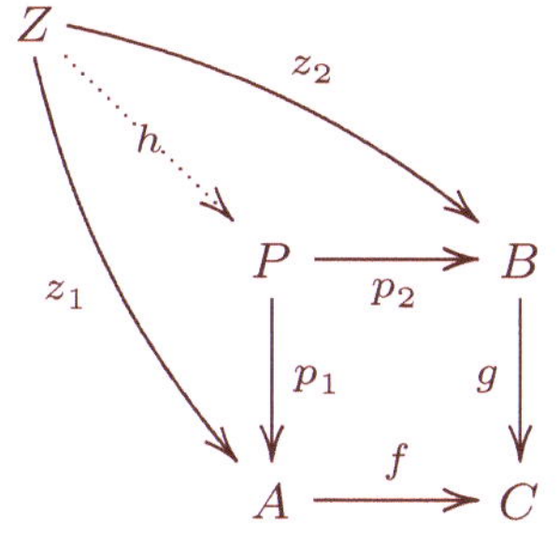

A pullback may be seen as a constrained product, being the constraint given by f and g, $f \circ p_1 = g \circ p_2$ as in Fig. 9.5.

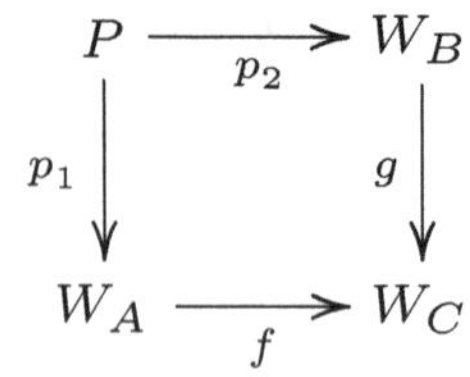

Fig. 9.5 The object P together with *arrows* p_1 and p_2 is a pullback in the category of mental objects C

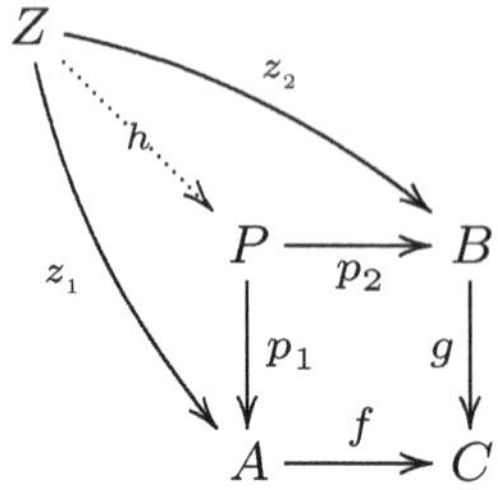

Fig. 9.6 The object P can be seen as a grid cell where *arrow* p_1 represents orientation and p_2 translation

9.3.3 Pullback and Grid Cell Formation

As it was shown in Chap. 8, the standard procedure to relate grid cells and place cells is problematic. One of the three parameters normally used to model grid cells refers to how accurate the representation outside the cell is. In doing so the quality of the system to work as a grid cell can be tuned at will by the modeler. This makes the model, to a certain extent, is unreliable because the modeler is introducing as an independent variable the expected behavior that supposedly the grid cell will provide. Here we show that the concept of pullback is useful in dealing with grid cells.

In Fig. 9.6, P can be seen as a grid cell where the projection morphisms p_1 and p_2 refer to the orienting and the spacing respectively. The morphisms f and g impose additional restrictions in the grid cell, like for example the constant value of those parameters all over the arena.

Chapter 10
Epilogue

The progress from an immature science to a mature one must pass through the construction of appropriate mechanisms with explanatory power. In brain science we must learn from the lessons geneticists have learnt in the hard way. The genome tells us nothing about how proteins fold and interact with each other, and the clinical benefits of the human genome project are far from clear.

Neuroscience is at a historic turning point. Two decades after the U.S. congress declared the 1990s the decade of the brain [446], brain research has accelerated at a vertiginous pace. Since then, basic science developments, together with technological achievements in brain imaging have lead neuroscience to a breathtaking momentum. The conceptual shift in neuroscience has been supported by discoveries like the understanding of the molecular implications in Alzheimer's disease, which has recast the concept of mental disease as a brain dysfunction that can be further conceptualized as a network malfunction. It is possible to monitor in vivo the developing brain, and large scale brain systems are being simulated in fully anatomical and physiological details. However, we do not have the answer to the big questions about learning, memory or intelligence. The global picture of brain functioning is still missing.

It is unreasonable to think that purely linear mathematics can be the killer tool to model the complex interactions that take place in the brain. We may indeed, need a new mathematical language for describing brain activity. The understanding of brain function and dynamics will require to integrate knowledge from a multidisciplinary perspective, setting the ground for an integrative neuroscience built upon scalable models at intercellular, cellular and molecular levels. New ways of thinking about integration are more necessary than ever in the "synthesis era in neuroscience [447]".

The problem of how cognition arises out of brain dynamics can be studied as the formation and evolution of a special kind of patterns, those that are relevant for cognition. What is missing is to formulate the general principles that govern the deployment of these different patterns on multiple scales. The immediate future of

J. Gómez-Ramirez, *A New Foundation for Representation in Cognitive and Brain Science*, Springer Series in Cognitive and Neural Systems 7, DOI 10.1007/978-94-007-7738-5_10,

neuroscience is unmistakably multidisciplinary. The challenge ahead is to foster the synthesis and integration of the data and models produced in the past two decades. The study of the structure-function problem in the brain requires a systemic theory based approach to attain a system level understanding that favours the construction of models that are not merely descriptive or phenomenological – curve-fitting data – but takes steps towards mechanistic models that provide real insights into critical parameters that control the system dynamics [40].

The new mathematical framework provided in this book is directed to set the necessary ground to build on models able to deal with multilevel interactions of coupled elements with non linear dynamics. In this vein, category theory is here used as an analytical tool for investigating the common properties of certain objects. The applications shown in this work, depict the suitability of Category Theory as a language for complex system modeling, and as sophisticated toolkit for mental theories. Relatively recent mathematical domains like category theory, or older ones, that were in the past set aside from the mathematician's agenda, have recently gained a renewed interest thanks to their application in biological systems e.g. heartbeat patterns modeled with groupoids [100]. The network-based approach is being pushed forward in different fields of biomedical research, however network biology is still in its infancy [159]. The hierarchical interdependency found in biological systems needs to be addressed taking into account the hierarchies and feedback loops among the network components. However, a purely graph theoretic approach based on topological analysis of networks is inefficient to account for the biological relevance of network motifs. The mathematical theory of categories provides a higher level of analysis compared to graph theory and can be effectively used to explore large scale network properties. Specifically, colimits – construction schemes that generalizes constructions such as disjoint unions, direct sums, coproducts, pushouts and direct limits – allow us to obtain universal classifiers of network topologies and mappings.

The examples presented in the book demonstrates that the study of brain connectivity patterns can be efficiently addressed with categorical concepts. This book, therefore, tries to takes steps towards characterizing the interplay between structure and function in brain networks, avoiding the limitations of standard approaches, and suggesting new pathways that may lead to the comprehension of the recurrent patterns and structures that are necessarily involved in any cognitive process. Furthermore, this work is intended to set the agenda of category theory as a valid methodology in cognitive and brain science. Although mathematical categories, at first sight, may seem too pure and universal in contrast with the spurious biological realm, where the particular prevails over the universal, it may lead to new and deeper insights into the structure and the representational power of the brain. In particular, we can identify the following five future developments of the work presented here:

1. To characterize the polychromous neural groups by a categorical property able to extend the Cohen-Grossberg-Hopfield [448] model (including delay).

2. To define patterns of connections that serve as building blocks of the networks, and colimits of patterns that are meaningful patterns or neural masses with representational power.
3. Investigate which are the general principles that rule the emergence of meaningful patterns in the brain. The mechanisms by which patterns are created and maintained, need to be tackled based on the mathematical formulation of terms like stability, robustness or autoregulation studied in Control Theory.
4. The hypothesis established in the book: co-operation of the grid fields gives rise to the colimit which is a place field, must be empirically validated with humans, analysing EEG data streams from the hippocampus and other brain areas involved.
5. Despite the diverse nature of episodic and semantic memory, it is possible to connect memories via categorical objects like product, pullback or colimit. Place and grid cells need to be tested in humans as possible building blocks for episodic memory.

In science, many problems are still unsolved, otherwise there would not be scientists, but technicians. And many questions that we want to solve today will never have an answer, and not necessarily because they are too complex to be solved, but because they will just disappear for irrelevant. In the end, brain function and behaviour depend upon the interaction of elements (cells, networks, brains). The neural and cognitive sciences, deal with complex components (assemblies of cells, brains) and out of necessity will follow a mathematical strand. Science moves in little steps or "one funeral at a time" as Planck caustically pointed out, but it also progresses with revolutionary discoveries and concepts that sweep away whole and entire edifices of thinking and replace them with new theories that explain more with less.

In the absence of a proper mathematical language for expressing brain function theories, the separation between the different levels of investigation relevant for brain function and dynamics will be progressively more acute and a consilience between them unattainable. The real value of this work for the future must reside in its capacity to foster the interest of either experimentalists and network theorists in validating and expanding the methodologies and theories exposed here.

References

1. Eliot TS (2001) Four quartets: quartet no. 2: east coker. Faber and Faber, London
2. Jeannerod M (1999) A dichotomous visual brain? Psyche 5. Retrieved October 19, 2013, from www.theassc.org/files/assc/2443.pdf
3. Sokal A, Bricmont J (1999) Fashionable nonsense: postmodern intellectuals' abuse of science, 1st edn. Picador, New York
4. Thompson DW (1992) On growth and form. Cambridge University Press, Cambridge
5. Kendrew J (1961) The three-dimensional structure of a protein molecule. Sci Am 205:96–110
6. Cull P (2007) The mathematical biophysics of nicolas rashevsky. Biosystems 88(3):178–184. BIOCOMP 2005: selected papers presented at the international conference – diffusion processes in neurobiology and subcellular biology, BIOCOMP2006: diffusion processes in neurobiology and subcellular biology
7. Gerstein MB, Bruce C, Rozowsky JS, Zheng D, Du J, Korbel JO, Emanuelsson O, Zhang ZD, Weissman S, Snyder M (2007) What is a gene, post-ENCODE? history and updated definition. Genome Res 17(6):669–681. PMID: 17567988
8. White JG, Southgate E, Thomson JN, Brenner S (1986) The structure of the nervous system of the nematode caenorhabditis elegans. Philos Trans R Soc Lond B Biol Sci 314(1165):1–340
9. Schrödinger E (1944) What is life? Mind and matter. Cambridge University Press, Cambridge
10. Barr M, Wells C (1997) This is biology: the science of the living world. Belknap Press of Harvard University Press, Cambridge
11. Wikipedia Action potential. http://en.wikipedia.org/wiki/Action_potential
12. Golubitsky M (2006) Symmetry and neuroscience. Bulletin of the AMS. American Mathematical Society (January 14)
13. Eilenberg S (Aug 2013) Page Version ID: 566981902
14. Mac Lane S (Aug 2013) Page Version ID: 550989403
15. Grothendieck A (Aug 2013) Page Version ID: 566771460
16. Lawvere W (Aug 2013) Page Version ID: 548025672
17. Healy J, Caudell T (2006) Ontologies and worlds in category theory: implications for neural systems. Axiomathes 16:165–214
18. Neuman Y, Nave O (2008) On the semio-mathematical nature of codes. Biosemiotics 1:99–111
19. Edelman G, Tononi G (2001) Consciousness: how matter becomes Imagination. Penguin, London
20. (2013) One and Three Chairs. June 2013. Page Version ID: 559080504
21. Ruth Lawson Otago Polytechnic (2007) Anatomy and physiology of animals relation btw sensory, relay & motor neurons. http://commons.wikimedia.org/wiki/File:Anatomy_and_physiology_of_animals_Relation_btw_sensory,_relay_%26_motor_neurons.jpg

J. Gómez-Ramirez, *A New Foundation for Representation in Cognitive and Brain Science*, Springer Series in Cognitive and Neural Systems 7, DOI 10.1007/978-94-007-7738-5,

22. Hubel DH, Wiesel TN (1977) Ferrier lecture. Functional architecture of macaque monkey visual cortex. Proc R Soc Lond Ser B. Containing papers of a Biological character Royal Society (Great Britain) 198(1130):1–59. PMID: 20635
23. Fuhs M, Touretzky D (2006) A spin glass model of path integration in rat medial entorhinal cortex. J Neurosci 26:4266–76
24. Taube J (2009) Head direction cells. Scholarpedia 4(12):1787
25. Mazzocchi F (2008) Complexity in biology. Exceeding the limits of reductionism and determinism using complexity theory. EMBO Rep 9(1):10–14. PMID: 18174892, PMCID: PMC2246621
26. Gómez-Ramirez J, Sanz R (2013) On the limitations of standard statistical modeling in biological systems: a full Bayesian approach for biology. Prog Biophys Mol Biol 113(1):80–91
27. Freeman WJ, Kozma R (2010) Freeman's mass action. Scholarpedia 5(1):8040
28. Carnap R (1967) The logical structure of the world. Pseudoproblems in philosophy. University of California Press, Berkeley
29. Kitano H (2002) Looking beyond the details: a rise in system-oriented approaches in genetics and molecular biology. Curr Genet 41(1):1–10. PMID: 12073094
30. Strange K (2005) The end of naive reductionism: rise of systems biology or renaissance of physiology? Am J Physiology Cell Physiol 288(5):C968–C974
31. Jacrot B (2006) Physique et biologie. Edp Sciences, Les Ulis
32. Taylor JG (1999) Towards the networks of the brain: from brain imaging to consciousness. Neural Netw 12(7–8):943–959. PMID: 12662638
33. Bower JM (2005) Looking for Newton: realistic modeling in modern biology. In: Bower JM, Beeman D (eds) Special issue on realistic neural modeling - Wam-Bamm '05 tutorials. Brains, minds and media, vol 1, bmm217
34. Noble D (2010) Biophysics and systems biology. Philos Trans R Soc A 368(1914):1125–1139
35. Brenner S (1999) Theoretical biology in the third millennium. Philos Trans R Soc Lond B 7:1963–1965
36. Smolin L (2003) The self-organization of space and time. Philos Trans Ser A Math Phys Eng Sci 361(1807):1081–8
37. Knight J (2002) Bridging the culture gap. Nature 419:244–246
38. Kuhn T (1962) The structure of scientific revolutions. University of Chicago Press, Chicago
39. Kuhn T (2000) The road since structure: philosophical essays: 1970-1993. University of Chicago Press, Chicago
40. Gatenby RA, Maini PK (2003) Mathematical oncology: cancer summed up. Nature 421(6921):321. PMID: 12540881
41. Polger T (2004) Natural minds. MIT, Cambridge
42. Bechtel W, Mundale J (1999) Multiple realizability revisited: linking cognitive and neural states. Philos Sci 66:175–207
43. Buzsaki G (2006) Rhythms of the brain. Oxford University Press, Oxford/New York
44. Nicolis G, Prigogine I (1977) Self-organization in nonequilibrium systems: from dissipative structures to order through fluctuations, 1st edn. Wiley, New York
45. Whishaw I, Vanderwolf C (1973) Hippocampal EEG and behavior: change in amplitude and frequency of RSA (theta rhythm) associated with spontaneous and learned movement patterns in rats and cats. Behav Biol 8(4):461–484
46. Pare D, Llinas R (1995) Conscious and pre-conscious processes as seen from the standpoint of sleep-waking cycle neurophysiology. Neuropsycholgia 33(9):1155–1168
47. Nunez PL (1995) Neocortical dynamics and human EEG rhythms. Oxford University Press, New York
48. Nunez P (2000) Toward a quantitative description of large-scale neocortical dynamic function and EEG. Behav Brain Sci 23(3):371–98
49. Chalmers D (1995) Facing up to the problem of consciousness. J Conscious Stud 2(3):200–219
50. Fell J (2004) Identifying neural correlates of consciousness: the state space approach. Conscious Cogn 13(4):709–729

51. Feldman J (2009) Ecological expected utility and the mythical neural code. Cogn Neurodyn 4(1):25–35
52. Gómez-Ramirez J (2012) Comments on aur's "from neuroelectrodynamics to thinking machines". Cogn Comput 4(4):563–565
53. Rosenblatt F (1958) The perceptron: a probabilistic model for information storage and organization in the brain. Cornell Aeronaut Lab Psychol Rev 65(6):386–408
54. Hodgkin A, Huxley A (1982) A quantitative description of membrane current and its application to conduction and excitation in nerve. J Physiol 117:500–544
55. Rall W (2009) Rall model. Scholarpedia 4(4):1369
56. De Schutter E (2008) Why are computational neuroscience and systems biology so separate? PLoS Comput Biol 4(5):e1000078
57. Solinas S, Maex R, De Schutter E (2006) Dendritic amplification of inhibitory postsynaptic potentials in a model Purkinje cell. Eur J Neurosci 23(4):1207–1218
58. Niebur E (2008) Neuronal cable theory. Scholarpedia 3(5):2674
59. Rall W (2011) Core conductor theory and cable properties of neurons. In: Comprehensive physiology. Wiley
60. Segev I, London M (2000) Untangling dendrites with quantitative models. Science 290(5492):744–750
61. Segev I (2006) What do dendrites and their synapses tell the neuron? J Neurophysiol 95(3):1295–1297
62. Gerstner W, Kistler W (2002) Spiking neuron models. Single neurons, populations, plasticity. Cambridge University Press, Cambridge/New York
63. Bressloff P (2008) A Green's function approach to analysing the effects of random synaptic background activity in a model neural network. J Phys A Math Gen 27(12):4097
64. Abbott L (1999) Lapique's introduction of the integrate-and-fire model neuron. Brain Res Bull 50(5-6):303–304
65. Dayan P, Abbott L (2002) Theoretical neuroscience: computational and mathematical modeling of neural systems. MIT, Cambridge MA
66. Deco G, Jirsa V, Robinson P, Breakspear M, Friston K (2008) The dynamic brain: From spiking neurons to neural masses and cortical fields. PLoS Comput Biol 4(8):23–37
67. Gerstner W, Naud R (2009) Neuroscience: how good are neuron models? Science 326(5951):379–380
68. Meunier C, Segev I (2002) Playing the devil's advocate: is the Hodgkin-Huxley model useful? Trends Neurosci 25(11):558–63
69. Herz AVM, Gollisch T, Machens CK, Jaeger D (2006) Modeling single-neuron dynamics and computations: a balance of detail and abstraction. Science 314(5796):80–85
70. Markram H (2006) The blue brain project. Nature Neurosci 7:153–158
71. Davie JT, Kole MHP, Letzkus JJ, Rancz EA, Spruston N, Stuart GJ, HÃusser M (2006) Dendritic patch-clamp recording. Nat Protoc 1(3):1235–1247
72. von der Malsburg C, Phillips WA, Singer W (eds) (2010) Dynamic coordination in the brain: from neurons to mind. MIT, Cambridge
73. McCulloch W, Pitts WH (1943) A logical calculus of the ideas immanent in nervous activity. Bull Math Biophys 5:115–133
74. Searle J (1980) Minds, brains and programs. Behav Brain Sci 3(3):417–424
75. Harnad S (1990) The symbol grounding problem. Physica D 42(10):335–346
76. Cohen PR (2006) If not turing's test, then what? AI Mag 26(4):335–346
77. Hoffman W (1998) Are neural networks a good model for cognition? Psycoloquy 9(12). Retrieved from http://www.cogsci.ecs.soton.ac.uk/cgi/psyc/newpsy?9.12
78. Fodor J, Pylyshyn Z (1983) Connectionism and cognitive architecture: a critical analysis. Cognition 28:3–71
79. Aizawa K (1995) The role of systematicity argument in classicism and connectionism. In: Ó Nuallaáin S, Mc Kevitt P, Mac Aogáin E (eds) Two sciences of mind: readings in cognitive science and consciousness. John Benjamims B.V, Amsterdam/Philadelphia
80. Eliasmith C (2007) Attractor network. Scholarpedia 2(10):1380

81. Izhikevich E (2006) Dynamical systems in neuroscience the geometry of excitability and bursting. MIT, Cambridge
82. Zhang D, Raichle ME (2010) Disease and the brain's dark energy. Nat Rev Neurol 6(1):15–28
83. Biswal B, Yetkin FZ, Haughton VM, Hyde JS (1995) Functional connectivity in the motor cortex of resting human brain using echo-planar MRI. Magn Reson Med 34(4):537–541. PMID: 8524021
84. Damoiseaux JS, Rombouts SARB, Barkhof F, Scheltens P, Stam CJ, Smith SM, Beckmann CF (2006) Consistent resting-state networks across healthy subjects. Proc Natl Acad Sci USA 103(37):13848–13853. PMID: 16945915
85. Fox M, Raichle M (2007) Spontaneous fluctuation in brain activity observed with functional magnetic resonance imaging. Nat Neurosci Rev 8:700–711
86. West G, Brown J, Enquist B (1997) A general model for the origin of allometric scaling laws in biology. Science 276:122–126
87. Castellano C, Fortunato S, Loreto V (2009) Statistical physics of social dynamics. Rev Mod Phys 81:591–646
88. Willems JC (1972) Dissipative dynamical systems part i: General theory. Arch Ration Mech Anal 45(5):321–351
89. David O, Friston K (2003) A neural mass model for MEG/EEG coupling and neuronal dynamics. NeuroImage 20:102–109
90. Haskell E, Nykamp DQ, Tranchina D (2001) A population density method for large-scale modeling of neuronal networks with realistic synaptic kinetics. Neurocomputing 38-40:627–632
91. Breakspear M, Stam C (2005) Dynamics of a neural system with a multiscale architecture. Philos Trans R Soc B (360):1051–1074
92. Freeman W (1975) Mass action in the nervous system. Academic, New York
93. Freeman W, Schneider W (1982) Changes in spatial patterns of rabbit olfactory EEG with conditioning to odors. Psychophysiology 19(1):44–56
94. Strogatz S, Stewart I (1993) Coupled oscillators and biological synchronization. Sci Am 12(1993):102–109
95. Frank T, Daffertshofer A, Peper C, Beek P, Haken H (2000) Towards a comprehensive theory of brain activity: coupled oscillator systems under external forces. Physica D 14:62–86
96. Hoppensteadt F, Izhikevich E (1997) Weakly connected neural networks. Springer, New York
97. Collins J, Stewart I (1999) Coupled nonlinear oscillators and the symmetries of animal gaits. J Nonlinear Sci 349:349–392
98. Golubitsky M, Stewart I, Buono PL, Collins J (1999) Symmetry in locomotor central pattern generators and animal gaits. Nature 401:693–695
99. Buono P, Golubitsky M (2001) Models of central pattern generators for quadruped locomotion. J Math Biol 42(4):291–326
100. Stewart I (2004) Network opportunities. Nature 427:601–604
101. Theunissen F, Roddey J, Stufflebeam S, Clague H, Miller J (1996) Information theoretic analysis of dynamical encoding by four primary sensory interneurons in the cricket cercal system. J Neurophysiol 75:1345–1359
102. Strogatz H (1994) Nonlinear dynamics and chaos. Addison Wesley, Reading
103. Abeles M (1991) Corticonics: neural circuits of the cerebral cortex. Cambridge University Press, Cambridge/New York
104. Manor Y, Rinzel J, Segev I, Yarom Y (1997) Low-amplitude oscillations in the inferior olive: a model based on electrical coupling of neurons with heterogeneous channel densities. J Neurophysiol 77:2736–2752
105. Steriade M, Domich L, Dakson G, Deschenes M (1987) The deafferented reticular thalamic nucleus generates spindle rhythmicity. J Neurophysiol 57:260–273
106. Caplan JB, Madsen JR, Schulze-Bonhage A, Aschenbrenner-Scheibe R, Newman EL, Kahana MJ (2003) Human theta oscillations related to sensorimotor integration and spatial learning. J Neurosci 23(11):4726–4736. PMID: 12805312

107. Garcia Dominguez L, Wennberg RA, Gaetz W, Cheyne D, Snead OC 3rd, Perez Velazquez JL (2005) Enhanced synchrony in epileptiform activity? Local versus distant phase synchronization in generalized seizures. J Neurosci 25(35):8077–8084. PMID: 16135765
108. Giaume C, McCarthy KD (1996) Control of gap-junctional communication in astrocytic networks. Trends Neurosci 19(8):319–325
109. Hutcheon B, Yarom Y (2000) Resonance, oscillation and the intrinsic frequency preferences of neurons. Trends Neurosci 23(5):216–222
110. Boehlen A, Heinemann U, Erchova I (2010) The range of intrinsic frequencies represented by medial entorhinal cortex stellate cells extends with age. J Neurosci 30(13):4585–9
111. Winfree AT (2001) The geometry of biological time, 2nd edn. Springer, New York
112. Acebrón JA, Bonilla LL, Pérez Vicente CJ, Ritort F, Spigler R (2005) The kuramoto model: a simple paradigm for synchronization phenomena. Rev Mod Phys 77(1):137–185
113. Ermentrout GB, Kopell N (1991) Multiple pulse interactions and averaging in systems of coupled neural oscillators. J Math Biol 29(3):195–217
114. Velazquez J (2007) Brain research: a perspective from the coupled oscillators field. NeuroQuantology 4(2):155–165
115. Buzsaki G (2010) Neural syntax: cell assemblies, synapsembles, and readers. Neuron 68(3):362–385
116. Freeman WJ (2008) A pseudo-equilibrium thermodynamic model of information processing in nonlinear brain dynamics. Neural Netw 21(2-3):257–265. PMID: 18249088
117. Prinz AA, Bucher D, Marder E (2004) Similar network activity from disparate circuit parameters. Nature Neurosci 7(12):1345–1352
118. Tsuda I (2001) Toward an interpretation of dynamic neural activity in terms of chaotic dynamical systems. Behav Brain Sci 24(05):793–810
119. Friston KJ (2001) Brain function, nonlinear coupling, and neuronal transients. Neurosci Rev J Bringing Neurobiol Neurol Psychiatry 7(5):406–418. PMID: 11597100
120. Harrison L, David O, Friston K (2005) Stochastic models of neuronal dynamics. Philos Trans R Soc B Biol Sci 360(1457):1075–1091. PMID: 16087449
121. Perez Velazquez J (2005) Brain, behaviour and mathematics: are we using the right approaches? Physica D 212:161–182
122. Griniasty M, Tsodyks M, Amit D (1999) Conversion of temporal correlations between stimuli to spatial correlations between attractors. In: Abbott LF, Sejnowski TJ (eds) Neural codes and distributed representations. MIT, Cambridge
123. Cossart R, Aromov D, Yuste R (2003) Attractor dynamics of network up states in the neocortex. Nature 423(5723):283–288
124. Hebb DO (2002) The organization of behavior: a neuropsychological theory, New edn. Psychology Press, Mahwah
125. Wills TJ, Lever C, Cacucci F, Burgess N, O'Keefe J (2005) Attractor dynamics in the hippocampal representation of the local environment. Science 308(5723):873–876
126. Lopes da Silva F et al (2003) Epilepsies as dynamical diseases of brain systems: basic models of the transition between normal and epileptic activity. Epilepsia 44(12):72–83
127. Babloyantz A, Destexhe A (1986) Low-dimensional chaos in an instance of epilepsy. Babloyantz 83(12):3513–3517
128. Titcombe MS et al (2001) Dynamics of Parkinsonian tremor during deep brain stimulation. Chaos 11:766–773
129. Prinz A, Bucher D, Marder (2000) Similar network activity from disparate circuit parameters. Nat Neurosci 7:1345–1352
130. Gómez-Ramirez J, Sanz R (2012) What the escherichia coli tells neurons about learning. In: Simeonov PL, Smith LS, Ehresmann AC (eds) Integral biomathics. Springer, Berlin/Heidelberg, pp 41–55
131. Mizumori SJY (2008) Hippocampal place fields: relevance to learning and memory, 1st edn. Oxford University Press, New York
132. Mac Cormac E, Stamenov M (1996) Fractals of brain, fractals of mind: in search of a symmetry bond. John Benjamins B.V., Amsterdam/Philadelphia

133. Nunn C (2008) From neurons to notions: brains, mind and meaning. Floris Books, Edinburgh
134. Haken H (1983) Synergetics: an introduction. Springer, Berlin
135. Haken H (2007) Synergetics. Scholarpedia 2(1):1400
136. Haken H (2006) Synergetics of brain function. Int J Psychophysiol 60(2):110–124
137. Haken H, Peper C, Beek P, Daffertshofer A (1996) A model for phase transitions in human hand movements during multifrequency tapping. Physica D Nonlinear Phenom 90(1–2): 179–196
138. Frank T, Daffertshofer A, Peper C, Beek P, Haken H (2000) Towards a comprehensive theory of brain activity: coupled oscillator systems under external forces. Physica D Nonlinear Phenom 144(1–2):62–86
139. Czaplicka A, Holyst JA, Sloot PMA (2013) Noise enhances information transfer in hierarchical networks. Sci Rep 3:1223
140. Pellionisz A, Llinás R (1985) Tensor network theory of the metaorganization of functional geometries in the CNS. Neuroscience 16:245–273
141. Makarenko V, Llinás R (1998) Experimentally determined chaotic phase synchronization in a neuronal system. Proc Natl Acad Sci USA 95:15747–15752
142. Leznik E, Makarenko V, Llinás R (2002) Electrotonically mediated oscillatory patterns in neuronal ensembles: an in vitro voltage-dependent dye imaging study in the inferior olive. J Neurosci 22(7):2804–2815
143. Roy S, Llinás R (2008) Dynamic geometry, brain function modeling, and consciousness. Prog Brain Res 168:133–144
144. Watts D, Strogatz S (1998) Collective dynamics of 'small-world' networks. Nature 393:244–442
145. Milo R, Shen-Orr S, Itzkovitz S, Kashtan N, Chklovskii D, Alon U (2002) Network motifs: simple building blocks of complex networks. Science 298(5594):824–827. PMID: 12399590
146. Newman MEJ (2003) The structure and function of complex networks. SIAM Rev 45(2):167–256
147. Hopfield J (1982) Neural networks and physical systems with emergent collective computational abilities. Proc Natl Acad Sci 79:2554–2558
148. Eliasmith C (2005) A unified approach to building and controlling spiking attractor networks. Neural Comput 17:1276–1314
149. Jaynes ET (1957) Information theory and statistical mechanics. Phys Rev 106:620–630
150. Jaynes ET (1957) Information theory and statistical mechanics. II. Phys Rev 108:171–190
151. Schneidman E, Berry MJ, Segev R, Bialek W (2006) Weak pairwise correlations imply strongly correlated network states in a neural population. Nature 440(7087):1007–1012. PMID: 16625187, PMCID: PMC1785327
152. Kaneko K (1992) Simulating spatiotemporal Chaos with coupled map lattices. In: Miyashita PDS, Imada PDM, Takayama PDH (eds) Computational approaches in condensed-matter physics. Springer, Berlin/Heidelberg, pp 260–271
153. Kaneko K, Tsuda I (2000) Complex systems: chaos and beyond, a constructive approach with applications in life sciences. Springer, Berlin/New York
154. Erdos P, Renyi A (1960) On the evolution of random graphs. Publ Math Inst Hung Acad Sci 5:17–61
155. Kozma R (2007) Neuropercolation. Scholarpedia 2(8):1360
156. Beggs JM (2007) Neuronal avalanche. Scholarpedia 2(1):1344
157. Beggs J, Plenz D (2003) Neuronal avalanches in neocortical circuits. J Neurosci 23(35):11167–11177
158. Breskin I et al (2006) Percolation in living neural networks. Phys Rev Lett 97(18):188102
159. Lee SY (ed) (2009) Systems biology and biotechnology of escherichia coli, 2009 edn. Springer, Dordrecht/London
160. Boden M (1972) What computers can't do: the limits of artificial intelligence. Harper & Row, New York
161. Boden M (2006) Mind as machine. Oxford University Press, New York

162. Lichtman JW, Livet J, Sanes JR (2008) A technicolour approach to the connectome. Nat Rev Neurosci 9:417–422
163. Lehrer J (2009) Neuroscience: making connections. Nature 457:524–527
164. Gowers T, Barrow-Green, Leader I (2008) The Princeton companion to mathematics. Princeton University Press, Princeton
165. Eilenberg S, MacLane S (1945) General theory of natural equivalences. Trans Am Math Soc 58(2):231
166. Barr M, Wells C (1999) Category theory for computing science. CRM Publications, Montreal.
167. Mac Lane S (1988) Concepts and categories in perspective. In: Duren P, A century of mathematics in America Part 3. American Mathematical Society, Providence 21:353–365
168. Pierce B (1991) Arrows, structures, and functors: the categorical imperative. MIT, Cambridge MA
169. Awodey S (2006) Category theory. Oxford University Press, Oxford/New York
170. Mac Lane S (1986) Mathematics form and function. Springer, New York
171. Mac Lane S (1998) Categories for the working mathematician. Springer, New York
172. Lawvere FW, Schanuel SH (1997) Conceptual mathematics: a first introduction to categories [...] [...]. Cambridge University Press, Cambridge
173. Arbib M, Manes E (1975) Arrows, structures, and functors: the categorical imperative. Academic, New York
174. Simmons H (2011) An introduction to category theory. Cambridge University Press, Cambridge/New York
175. Rydeheard DE, Burstall (1988) Computational category theory, 1st edn. Prentice Hall, New York
176. Borceux F (2008) Handbook of categorical algebra: volume 1, Basic category theory, 1st edn. Cambridge University Press, Cambridge
177. Pareigis B (1970) Categories and functors, 1st edn. Academic, New York
178. Schalk A, Simmons H (2005) Introduction to category theory in for easy movements. www.cs.man.ac.uk/mfg
179. Van Heijenoort J (1977) From Frege to Gödel: a source book in mathematical logic, 1879–1931. Harvard University Press, Cambridge/London
180. Macnamara J (1994) The logical foundations of cognition. Oxford University Press, New York
181. Hoffman W (2012) The dialectics of mind. J Mind Theory 1(1):1–27
182. Philips S, Wilson W, Hadford G (2009) What do transitive inference and class inclusion have in common? Categorical (co)products and cognitive development. PLoS Comput Biol 5(12):e1000599
183. Brown R, Porter T (2003) Category theory and higher dimensional algebra: potential descriptive tools in neuroscience. In: Singh N (ed) Proceedings of the international conference on theoretical neurobiology, Delhi, 2003. National Brain Research Centre, conference proceedings, vol 1, pp 80–92
184. Ehresmann A, Vanbremeersch JP (2007) Memory evolutive systems; hierarchy, emergence, cognition. Elsevier, Amsterdam/Boston/Paris
185. Zeeman EC (1962) The topology of the brain and visual perception. Prentice-Hall, New York
186. Spivak DI, Kent RE (2012) Ologs: a categorical framework for knowledge representation. PLoS ONE 7(1):e24274
187. Fidaeiro JL (2004) Categories for software engineering. Springer, Berlin Heidelberg
188. Haskell (programming language) (2013). Page Version ID: 551056925. http://en.wikipedia.org/wiki/Haskell_%28programming_language%29
189. Scala (programming language) (2013). Page Version ID: 551000017. http://en.wikipedia.org/wiki/Scala_%28programming_language%29
190. Clojure (2013). Page Version ID: 548200920. http://en.wikipedia.org/wiki/Clojure
191. Kary M (2009) (Math, science, ?). Axiomathes 19:321–339
192. Simeonov PL (2010) Integral biomathics: a post-Newtonian view into the logos of bios. Prog Biophys Mol Biol 102(2-3):85–121. PMID: 20146929

193. Rosen R (2006) Autobiographical reminiscences of rober rosen. Axiomathes 16:1–23
194. Letelier JC, Soto-Andrade J, Abarzúa FG, Cornish-Bowden A, Cárdenas ML (2006) Organizational invariance and metabolic closure: analysis in terms of (m,r) systems. J Theor Biol 238(4):949–961
195. Arzi-Gonczarowski Z, Lehmann DJ (1998) From environments to representations: a mathematical theory of artificial perceptions. Artif Intell 102:187–247
196. Rashevsky N (1967) Organismic sets: outline of a general theory of biological and social organisms. Bull Math Biophys 29(1):139–152
197. Rashevsky N (1968) Organismic sets: II. Some general considerations. Bull Math Biophys 30(1):163–174
198. Healy M (2000) Category theory applied to neural modeling and graphical representations. In: Proceedings of IJCNN 2000, Como. IEEE
199. Neuman YN, Nave O (2008) A mathematical theory of sign-mediated concept formation. Appl Math Comput 201:72–81
200. Shannon CE (1948) A mathematical theory of communication. Bell Syst Tech J 27:370–423
201. Tamarin RH (2002) Principles of genetics. McGraw-Hill, New York
202. Gomez J, Sanz R (2009) To cognize is to categorize revisited: category theory is where mathematics meets biology. AAAI fall symposium series, Washigton DC (Online)
203. Rosch E, Mervis CB (1975) Family resemblances: studies in the internal structure of categories. Cogn Psychol 7:573–605
204. Rosch E (1978) Cognition and categorization. Erlbaum, Hillsdale
205. Gómez-Ramirez J (2013) Naturalizing epistemology for autonomous systems. In: Talmont-Kaminski K, Milkowski M (eds) Regarding the mind, naturally: naturalist approaches to the sciences of the mental. Cambridge Scholars Publishing, Newcastle upon Tyne, pp 232–247
206. Philips S, Wilson WH (2010) Categorial compositionality: a category theory explanation for the systematicity of human cognition. PLoS Comput Biol 6(7):14
207. Dalton TC (2005) Challenging philosophical assumptions about mind. Trends Cogn Sci 9(8):365–366
208. Steels L (2006) Experiments on the emergence of human communication. Trends Cogn Sci 10(8):347–349
209. Ramus F (2006) Genes, brain, and cognition: a roadmap for the cognitive scientist. Cognition 101(2):247–269
210. Banerjee K, Bloom P (2013) Would Tarzan believe in God? Conditions for the emergence of religious belief. Trends Cogn Sci 17(1):7–8
211. Shiffrin RM (2010) Perspectives on modeling in cognitive science. Top Cogn Sci 2(4):736–750
212. Koblitz N (1988) A tale of three equations; or the emperors have no clothes. Math Intell 10(1):4–10
213. Huntington S (1968) Political order in changing societies. Yale University Press, New Haven
214. Brown J (1999) Philosophy of mathematics: an introduction to the world of proofs and pictures. Routledge, London/New York
215. Koblitz N (1981) Mathematics as propaganda. In: Steen DLA (ed) Mathematics tomorrow. Springer, New York, pp 111–120
216. Wille R, Wille U (1993) On the controversy over huntington's equations: when are such equations meaningful? Math Soc Sci 25(2):173–180
217. Changeaux JP, Connes A (2000) Matière à pensée. Odile Jacob, Paris
218. Nagel E (1979) The structure of science: problems in the logic of scientific explanation. Hacket Publishing, Indianapolis
219. Bunge M (1969) La Investigación Científica. Ariel España, Barcelona
220. Bunge M (2003) Emergence and convergence: qualitative novelty and the unity of knowledge. University of Toronto Press, Toronto
221. Bunge M (1979) Some topical problems in biophilosophy. J Soc Biol Syst 2(2):155–172
222. Ahn AC, Tewari M, Poon CS, Phillips RS (2006) The limits of reductionism in medicine: could systems biology offer an alternative? PLoS Med 3(6):e208

223. Gómez-Ramirez J, Wu J (2012) A new vision for biomedicine: a systems approach. In: 2012 ICME international conference on complex medical engineering (CME), Kobe, pp 479–484
224. Courant R, Robbins H, Stewart I (eds) (1996) What is mathematics? An elementary approach to ideas and methods. Oxford University Press, New York
225. Bohm D (2002) The essential David Bohm. Routledge, London
226. Dehaene E (2007) A few steps toward a science of mental life. Mind Brain Educ 1(1):28–47
227. Luce R (1995) Four tensions concerning mathematical modeling in psychology. Annu Rev Psychol 46:1–26
228. van Dalen D (1997) How connected is the intuitionistic continuum? J Symb Log 62(4):1147–1150
229. Fodor J (1983) The modularity of mind. MIT, Cambridge
230. Shannon CE, Weaver W (1949) The mathematical theory of communication, vol 27. University of Illinois Press, Urbana
231. Kandel E (2000) Principles of neural science. McGraw-Hill, New York
232. Andersen P, Morris R, Amaral D, Bliss T, O'Keefe J (eds) (2006) The Hippocampus book, 1st edn. Oxford University Press, New York
233. Squire L, Berg D, Bloom FE, Lac Sd, Ghosh A, Spitzer NC (eds) (2012) Fundamental neuroscience, 4th edn. Academic, Oxford
234. Moore GP, Perkel DH, Segundo JP (1966) Statistical analysis and functional interpretation of neuronal spike data. Annu Rev Physiol 28(1):493–522. PMID: 5323104
235. Borst A, Theunissen FE (1999) Information theory and neural coding. Nat Neurosci 2(11):947–957
236. Rolls E (2007) Memory, attention, and decision-making: a unifying computational neuroscience approach, 1st edn. Oxford University Press, Oxford
237. Wang P (2007) Three fundamental misconceptions of artificial intelligence. J Exp Theor Artif Intell 19(3):249–268
238. Bays T (2012) Skolem's paradox. In: Zalta EN (ed) The Stanford encyclopedia of philosophy, winter 2012. http://plato.stanford.edu/archives/win2012/entries/paradox-skolem/
239. Longo G (1997) Mathematical intelligence, infinity and machines: beyond the godelitis. J Conscious Stud 6:191–214
240. Dive LL (2003) An epistemic structuralist account of mathematical knowledge. PhD thesis, Department of Philosophy, Faculty of Arts, University of Sydney
241. Petitot J et al (1999) Naturalizing phenomenology: issues in contemporary phenomenology and cognitive science. Stanford University Press, Stanford
242. Arbib MA (1972) The metaphorical brain: introduction to cybernetics as artificial intelligence and brain theory, 1st edn. Wiley, New York
243. Pulvermüller F (1996) How and where are words represented and processed in the brain? In: International workshop on language, brain and verbal behavior: neurobiological aspects of linguistic capacities and language processing, Barcelona. Institut d'Estudis Catalans, pp 63–75
244. Feldman J (2008) From molecule to metaphor: a neural theory of language, 1st edn. A Bradford Book. The MIT Pess, Cambridge MA
245. Feldman J, Lakoff G, Bailey D, Narayanan S, Regier T, Stolcke A (1996) L0 – the first five years of an automated language acquisition project. Artif Intell Rev 10(1-2):103–129
246. Salthe S (2004) The spontaneous origin of new levels in a scalar hierarchy. Entropy 6:327–343
247. Salthe S (2009) A hierarchical framework for levels of reality: understanding through representation. Axiomathes (19):87–99
248. Baas N, Emmeche C (1997) On emergence and explanation. Intellectica (25):67–83
249. Putnam H (1981) Reason, truth and history. Cambridge University Press, Cambridge
250. Dummett M (2006) Thought and reality. Oxford University Press, New York
251. Holland J (1999) Emergence: from chaos to order. Basic Books, New York
252. Chalmers DJ (2007) Consciousness and its place in nature. In: Stich SP, Warfield TA (eds) The Blackwell guide to philosophy of mind. Blackwell, Malden, pp 102–142

253. Mack G (1996) Gauge theory of things alive: universal dynamics as a tool in parallel computing. Prog Theor Phys Suppl 122:201–212
254. Thom R (1988) Esquisse d'une sémiophysique. InterEditions, Paris
255. Mccarthy J, Hayes PJ (1969) Some philosophical problems from the standpoint of artificial intelligence. In: Michie D, Meltzer B (eds) Machine intelligence. Edinburgh University Press, Edinburgh, pp 463–502
256. Shanahan M (1997) Solving the frame problem: a mathematical investigation of the common sense law of inertia. MIT, Cambridge
257. Baianu IC et al (2007) Categorical ontology of complex spacetime structures: the emergence of life and human consciousness. Axiomathes (17):223–352
258. James W (1890) The principles of psychology. Henry Holt, New York
259. Rosen R (1985) Organisms as causal systems which are not mechanisms: an essay into the nature of complexity'. In: Rosen R (ed) Theoretical biology and complexity. Academic, Orlando, pp 165–203
260. Ehresmann AC, Vanbremeersch JP (2006) The memory evolutive systems as a model of Rosenâs organisms â (metabolic, replication) systems. Axiomathes 16(1–2):137–154
261. Pawlowitsch C (2008) Why evolution does not always lead to an optimal signaling system. Games Econ Behav 63:203–226
262. Wagner A (2007) Robustness and evolvability in living systems. Princeton University Press, Princeton/Woodstock
263. Alon U, Surette MG, Barkai N, Leibler S (1999) Robustness in bacterial chemotaxis. Nature 397(6715):168–171
264. Wagner A (2000) Robustness against mutations in genetic networks of yeast. Nat Genet 24(4):355–361. PMID: 10742097
265. Ay N, Krakauer DC (2007) Geometric robustness theory and biological networks. Theory Biosci (Theorie in den Biowissenschaften) 125:93–121
266. Tononi O Gand Sporns, Edelman G (1999) Measures of degeneracy and redundancy in biological networks. Proc Natl Acsd Sci 96(6):3257–3262
267. Holland JH (1999) Emergence: from chaos to order. Perseus Books, Reading
268. Zee A (2007) Fearful symmetry: the search for beauty in modern physics. Princeton Science Library, Princeton
269. Derrida J, Roudinesco E (2004) For what tomorrow: a dialogue. Stanford University Press, Stanford
270. Livingston E (1999) Cultures of proving. Soc Stud Sci 29(6):867–888
271. Miller GA (2007) Charge densities of the neutron and proton. Phys Rev Lett 99(11):112001
272. Elasser W (1958) The physical fundation of biology. Pergamon Press, London
273. Katz M (1986) Templets and complet patterns. Cambridge University Press, Cambridge MA
274. Arbib MA, Hesse MB (1986) The construction of reality. Cambridge University Press, Cambridge/New York
275. Corry L (2004) Modern algebra and the rise of mathematical structures, 2nd edn. Birkhauser, Basel/Boston
276. Shapiro S (1996) Space, number and structure: a tale of two debates. Philos Math 4(3):148–173
277. Skyttner L (2006) General systems theory: problems, perspectives, practice. World Scientific, Singapore
278. Alon U (2006) An introduction to systems biology: design principles of biological circuits, 1st edn. Chapman and Hall/CRC, Boca Raton/London
279. Palsson BO (2006) Systems biology: properties of reconstructed networks, 1st edn. Cambridge University Press, Cambridge/New York
280. Mac Lane S (2006) Category theory as a framework for an in re interpretation of mathematical structuralism. Cah Topol Geom 21:163–179
281. Landry E (1998) Category theory: the language of mathematics. Philos Sci 66:14–27
282. Baez JC (1997) An introduction to n-categories. In: Moggi E, Rosolini G (eds) Category theory and computer science. No. 1290 in lecture notes in computer science. Springer, Berlin/Heidelberg, pp 1–33

283. Shapiro S (1996) Structure in mathematics and logic: a categorical perspective. Philos Math 4(4):209–237
284. Dragoi G, Tonegawa S (2011) Preplay of future place cell sequences by hippocampal cellular assemblies. Nature 469(7330):397–401
285. Goodrich R (1988) An agent-based conception of models and scientific representation. Br J Aesthet 28(1):48–58
286. Swoyer C (1991) Structural representation and surrogative reasoning. Synth 87(3):449–508
287. Teller P (2001) Twilight of the perfect model. Erkenntnis 55(3):393–415
288. Suarez M (2003) Scientific representation: against similarity and isomorphism. Int Stud Philos Sci 17(3):225–244
289. Carver C (2006) When mechanistic models explain. Synthese 153:355–376
290. Woodward J (2005) Making things happen: a theory of causal explanation. Oxford University Press, New York/Oxford
291. Giere RN (1990) Explaining science: a cognitive approach. University of Chicago Press, Chicago
292. Giere RN (1999) Science without laws. University of Chicago Press, Chicago
293. Giere R (2010) An agent-based conception of models and scientific representation. Synthese 172(2):269–281
294. Shapiro S (2000) Thinking about mathematics. Oxford University Press, New York
295. Shapiro S (2005) Oxford handbook of the philosophy of mathematics and logic. Oxford University Press, Oxford/New York
296. Birkhoff G, Mac Lane S (1946) A survey of modern algebra. Macmillan, New York
297. Geroch R (1985) Mathematical physics. University of Chicago Press, Chicago
298. Bonsignorio F (2010) On the stochastic stability and observability of controlled serial kinematic chains. In: Proceedings of the ASME 2010 10th biennial conference on engineering systems design and analysis ESDA2010, Turkey
299. von Helmholtz H (2005) Treatise on physiological optics. Dover, Mineola
300. Poincaré H (1952) Science and hypothesis. Dover, New York
301. Grant G (2007) How the 1906 Nobel Prize in physiology or medicine was shared between Golgi and Cajal. Brain Res Rev 55(2):490–498
302. Shepard RN, Chipman S (1970) Second-order isomorphism of internal representations: shapes of states. Cogn Psychol 1(1):1–17
303. Edelman S (1998) Representation is representation of similarities. Behav Brain Sci 21(4):449–467; discussion 467–498. PMID: 10097019
304. Churchland P (1986) Neurophilosophy toward a unified science of the mind-brain. MIT, Cambridge
305. Wiltschko W, Wiltschko R (1996) Collective dynamics of 'small-world' networks. J Exp Biol 199(1):29–38
306. Gray H, Standring S (2008) Gray's anatomy: the anatomical basis of clinical practice, 40th edn. Churchill-Livingstone, Elsevier
307. Yildirim FB, Sarikcioglu L (2007) Marie Jean Pierre Flourens (1794–1867): an extraordinary scientist of his time. J Neurol Neurosurg Psychiatry 78(8):852. 17635978, PMCID: PMC2117745
308. York GK, Steinberg DA (2011) Hughlings Jackson's neurological ideas. Brain 134(10):3106–3113. PMID: 21903729
309. Finger S (2001) Origins of neuroscience: a history of explorations into brain function. Oxford University Press, New York
310. Lashley K, Clark G (1946) The cytoarchitecture of the cerebral cortex of Ateles: a critical examination of the architectoic studies. J Comp Neurol 85:223–305
311. Wernicke C (1908) The symptom-complex of aphasia. Appleton, New York
312. Dykes RW, Ruest A (1986) What makes a map in somatosensory cortex? In: Jones EG, Peters A (eds) Sensory-motor areas and aspects of cortical connectivity. No. 5 in cerebral cortex, Springer, Boston, pp 1–29

313. Smith CUM (2008) Elementary units of cortical activity? The rise and fall of the cortical column. In: 12th annual meeting of the international society for the history of the neurosciences, Los Angeles
314. Mountcastle V (1997) The columnar organization of the cerebral cortex. Brain 120:701–722
315. Casanova MF (ed) (2005) Neocortical modularity and the cell minicolumn. Nova Biomedical Books, New York
316. Horton J, Adams D (2005) The cortical column: a structure without a function. Philos Trans R Soc B 360(1456):837–862
317. Blasdel G, Salama G (1986) Voltage-sensitive dyes reveal a modular organization in monkey striate cortex. Nature 321:579–585
318. Rakic P (1995) Radial versus tangential migration of neuronal clones in the developing cerebral cortex. Proc Natl Acad USA Sci 92(11):323–327
319. Rakic P (2000) Radial unit hypothesis of neocortical expansion. Novartis Found Symp 228:30–42; discussion 42–52. PMID: 10929315
320. Bressler S, Tognoli E (2006) Operational principles of neurocognitive networks. Int J Psychophysiol 60:139–148
321. Fuster J (2000) The module: crisis of a paradigm (book review, "the new cognitive neurosciences" 2nd edn, M.S. Gazzaniga, editor-in-chief, MIT Press). Neuron (26):51–53
322. Bechtel W (2001) The compatibility of complex systems and reduction: a case analysis of memory research. Minds Mach 11:83–502
323. Fiehn O (2001) Combining genomics, metabolome analysis, and biochemical modelling to understand metabolic networks. Comp Funct Genomics 2(3):155–168. PMID: 18628911
324. White J et al (1983) Factors that determine connectivity in the nervous system of Caenorhabditis elegans. Cold Spring Harb Symp Quant Biol 48(Pt 2):633–640
325. Smith J, Ellenberger H, Ballanyi K, Richter D, Feldman J (1991) Pre-Botzinger complex: a brainstem region that may generate respiratory rhythm in mammals. Science 254(5032):726–729
326. Horwitz B (2003) The elusive concept of brain connectivity. NeuroImage 19(2):466–470
327. Rodriguez P (2009) Neural decoding of goal locations in spatial navigation in humans with fMRI. Human Brain Mapp 31(3):3391–3397
328. Sharp D, Awad M, Warren J, Wise R, Vigliocco G, Scott S (2009) The neural response to changing semantic and perceptual complexity during language processing. Human Brain Mapp 31(3):365–377
329. Summerfield C, Egner T, Mangels J, Hirsch J (2006) Mistaking a house for a face: neural correlates of misperception in healthy humans. Cereb Cortex 16:500–508
330. Marian V, Shildkrot Y, Blumenfeld H, Kaushanskaya M, Faroqi-Shah Y, Hirsch J (1996) Cortical activation during word processing in late bilinguals: similarities and differences as revealed by functional magnetic resonance imaging. J Clin Exp Neuropsychol 29(3):247–265
331. Friston KJ (2009) Modalities, modes, and models in functional neuroimaging. Science 326(5951):399–403. PMID: 19833961
332. Friston KJ, Price CJ (2011) Modules and brain mapping. Cogn Neuropsychol 28(3–4): 241–250. PMID: 21416411, PMCID: PMC3335279
333. Muniz de Rezende A (1975) Le point de départ dans la philosophie de merleau-ponty. Revue Philosophique de Louvain 73(19):451–480
334. Bakker B (2005) The concept of circular causality should be discarded. Commentary on marc D. Lewis: Bridging emotion theory and neurobiology through dynamic system modeling. Behav Brain Sci 28:195–196
335. Granger CWJ (1969) Investigating causal relations by econometric models and cross-spectral methods. Econometrica 37(3):424–38
336. Seth AK (2008) Causal networks in simulated neural systems. Cogn Neurodyn 2(1):49–64
337. Fuster JM (2005) Cortex and mind: unifying cognition, 1st edn. Oxford University Press, Oxford
338. Bressler SL (2007) The formation of global neurocognitive state. In: Perlovsky LI, Kozma R (eds) Neurodynamics of cognition and consciousness, understanding complex systems. Springer, Berlin/Heidelberg, pp 61–72

339. Freeman W (2000) Neurodynamics: an exploration in mesoscopic brain dynamics, 1st edn. Springer, London/New York
340. Goebel R, Linden DE, Lanfermann H, Zanella FE, Singer W (1998) Functional imaging of mirror and inverse reading reveals separate coactivated networks for oculomotion and spatial transformations. Neuroreport 9(4):713–719. PMID: 9559944
341. Cordes D, Haughton VM, Arfanakis K, Carew JD, Turski PA, Moritz CH, Quigley MA, Meyerand ME (2001) Frequencies contributing to functional connectivity in the cerebral cortex in "resting-state" data. AJNR Am J Neuroradiol 22(7):1326–1333. PMID: 11498421
342. Friston KJ, Frith CD, Liddle PF, Frackowiak RS (1993) Functional connectivity: the principal-component analysis of large (PET) data sets. J Cereb Blood Flow Metab 13(1):5–14. PMID: 8417010
343. Comon P (1994) Independent component analysis, a new concept? Signal Process 36(3):287–314
344. Bell AJ, Sejnowski TJ (1995) An information-maximization approach to blind separation and blind deconvolution. Neural Comput 7(6):1129–1159. PMID: 7584893
345. Hyvarinen A, Oja E (2000) Independent component analysis: algorithms and applications. Neural Netw 13(4-5):411–430. PMID: 10946390
346. Stone JV (2002) Independent component analysis: an introduction. Trends Cogn Sci 6(2):59–64
347. McKeown MJ, Hansen LK, Sejnowsk TJ (2003) Independent component analysis of functional MRI: what is signal and what is noise? Curr Opin Neurobiol 13(5):620–629. PMID: 14630228
348. van de Ven VG, Formisano E, Prvulovic D, Roeder CH, Linden DEJ (2004) Functional connectivity as revealed by spatial independent component analysis of fMRI measurements during rest. Human Brain Mapp 22(3):165–178. PMID: 15195284
349. Beckmann CF, DeLuca M, Devlin JT, Smith SM (2005) Investigations into resting-state connectivity using independent component analysis. Philos Trans R Soc Lond Ser B Biol Sci 360(1457):1001–1013. PMID: 16087444
350. Friston KJ (1998) Modes or models: a critique on independent component analysis for fMRI. Trends Cogn Sci 2(10):373–375
351. He Y, Evans A (2010) Graph theoretical modeling of brain connectivity. Curr Opin Neurol 23(4):341–350. PMID: 20581686
352. Erdös P, Rényi A (1960) On the evolution of random graphs. In: Publ Math Inst Hung Acad Sci 5:17–61
353. Schöner G, Kelso JA (1988) Dynamic pattern generation in behavioral and neural systems. Science (New York, NY) 239(4847):1513–1520. PMID: 3281253
354. Kelso JAS (1995) Dynamic patterns: the self-organization of brain and behavior [...] [...]. MIT, Cambridge
355. Kelso JAS (2012) Multistability and metastability: understanding dynamic coordination in the brain. Philos Trans R Soc B Biol Sci 367(1591):906–918. PMID: 22371613, PMCID: PMC3282307
356. Hartwell LH, Hopfield JJ, Leibler S, Murray AW (1999) From molecular to modular cell biology. Nature 402(6761 Suppl):C47–C52. PMID: 10591225
357. Ravasz E, Somera AL, Mongru DA, Oltvai ZN, Barabási AL (2002) Hierarchical organization of modularity in metabolic networks. Science 297(5586):1551–1555. PMID: 12202830
358. Humphries M, Gurney K, Prescott T (2006) The brainstem reticular formation is a small-world, not scale-free, network. Proc R Soc B Biol Sci 273(1585):503–511. PMID: 16615219, PMCID: PMC1560205
359. Vaessen MJ, Hofman PAM, Tijssen HN, Aldenkamp AP, Jansen JFA, Backes WH (2010) The effect and reproducibility of different clinical DTI gradient sets on small world brain connectivity measures. NeuroImage 51(3):1106–1116. PMID: 20226864
360. Honey CJ, Thivierge JP, Sporns O (2010) Can structure predict function in the human brain? NeuroImage 52(3):766–776. PMID: 20116438

361. Tononi G, Sporns O, Edelman GM (1994) A measure for brain complexity: relating functional segregation and integration in the nervous system. Proc Natl Acad Sci 91(11):5033–5037
362. Sporns O, Tononi G, Edelman GM (1991) Modeling perceptual grouping and figure-ground segregation by means of active reentrant connections. Proc Natl Acad Sci USA 88(1):129–133. PMID: 1986358
363. Srinivasan R, Russell DP, Edelman GM, Tononi G (1999) Increased synchronization of neuromagnetic responses during conscious perception. J Neurosci 19(13):5435–5448. PMID: 10377353
364. Varela F, Lachaux JP, Rodriguez E, Martinerie J (2001) The brainweb: phase synchronization and large-scale integration. Nat Rev Neurosci 2(4):229–239
365. Ermentrout B (1998) Neural networks as spatio-temporal pattern-forming systems. Rep Prog Phys 61(4):353
366. Achard S, Salvador R, Whitcher B, Suckling J, Bullmore E (2006) A resilient, low-frequency, small-world human brain functional network with highly connected association cortical hubs. J Neurosci 26(1):63–72. PMID: 16399673
367. Salvador R, Suckling J, Coleman MR, Pickard JD, Menon D, Bullmore E (2005) Neurophysiological architecture of functional magnetic resonance images of human brain. Cereb Cortex 15(9):1332–1342. PMID: 15635061
368. Salvador R, Anguera M, Gomar JJ, Bullmore ET, Pomarol-Clotet E (2010) Conditional mutual information maps as descriptors of net connectivity levels in the brain. Front Neuroinform 4:115
369. Barabasi AL, Albert R (1999) Emergence of scaling in random networks. Science 286(5439):509–512
370. Clauset A, Shalizi CR, Newman MEJ (2009) Power-law distributions in empirical data. SIAM Rev 51(4):661–703
371. Stumpf MPH, Porter MA (2012) Critical truths about power laws. Science 335(6069):665–666. PMID: 22323807
372. Wasserman S, Faust K (1994) Social network analysis: methods and applications. Cambridge University Press, Cambridge/New York
373. Latora V, Marchiori M (2001) Efficient behavior of small-world networks. Phys Rev Lett 87(19):198701
374. Supekar K, Musen M, Menon V (2009) Development of large-scale functional brain networks in children. PLoS Biol 7(7):e1000157
375. Travers J, Milgram S (1969) An experimental study of the small world problem. Sociometry 32(4):425
376. Albert R, Jeong H, Barabási AL (1999) Internet: diameter of the world-wide web. Nature 401(6749):130–131
377. Schmitt JE, Lenroot RK, Wallace GL, Ordaz S, Taylor KN, Kabani N, Greenstein D, Lerch JP, Kendler KS, Neale MC, Giedd JN (2008) Identification of genetically mediated cortical networks: a multivariate study of pediatric twins and siblings. Cereb Cortex (New York, NY: 1991) 18(8):1737–1747. PMID: 18234689
378. He Y, Chen ZJ, Evans AC (2007) Small-world anatomical networks in the human brain revealed by cortical thickness from MRI. Cereb Cortex (New York, NY: 1991) 17(10):2407–2419. PMID: 17204824
379. Watts DJ, Strogatz SH (1998) Collective dynamics of small-world networks. Nature 393(6684):440–442
380. Kleinberg JM (2000) Navigation in a small world. Nature 406(6798):845–845
381. Sporns O, Zwi JD (2004) The small world of the cerebral cortex. Neuroinformatics 2(2): 145–162. PMID: 15319512
382. Yu S, Huang D, Singer W, Nikolic D (2008) A small world of neuronal synchrony. Cereb Cortex (New York, NY: 1991) 18(12):2891–2901. PMID: 18400792
383. Bassett DS, Bullmore E (2006) Small-world brain networks. Neuroscientist 12(6):512–523
384. Supekar K, Menon V, Rubin D, Musen M, Greicius MD (2008) Network analysis of intrinsic functional brain connectivity in Alzheimer's disease. PLoS Comput Biol 4(6):e1000100. PMID: 18584043, PMCID: PMC2435273

385. Liu Y, Liang M, Zhou Y, He Y, Hao Y, Song M, Yu C, Liu H, Liu Z, Jiang T (2008) Disrupted small-world networks in schizophrenia. Brain J Neurol 131(Pt 4):945–961. PMID: 18299296
386. Liao W, Zhang Z, Pan Z, Mantini D, Ding J, Duan X, Luo C, Lu G, Chen H (2010) Altered functional connectivity and small-world in mesial temporal lobe epilepsy. PLoS ONE 5(1):e8525
387. Costa LF, Rodrigues F, Travieso G, Boas V (2006) Characterization of complex networks: a survey of measurements. Adv Phys 56(1):167–242
388. Costa LF, Silva FN (2006) Hierarchical characterization of complex networks. J Stat Phys 125(4):841–872
389. Whitacre J, Bender A (2010) Degeneracy: a design principle for achieving robustness and evolvability. J Theor Biol 263(1):143–153. PMID: 19925810
390. Stevens AA, Tappon SC, Garg A, Fair DA (2012) Functional brain network modularity captures inter- and intra-individual variation in working memory capacity. PLoS ONE 7(1):e30468
391. Bullmore E, Sporns O (2012) The economy of brain network organization. Nat Rev Neurosci 13(5):336–349
392. Bullmore E, Sporns O (2009) Complex brain networks: graph theoretical analysis of structural and functional systems. Nat Rev Neurosci 10(3):186–198. PMID: 19190637
393. Vaessen MJ, Hofman PAM, Tijssen HN, Aldenkamp AP, Jansen JFA, Backes WH (2010) The effect and reproducibility of different clinical DTI gradient sets on small world brain connectivity measures. NeuroImage 51(3):1106–1116. PMID: 20226864
394. Deuker L, Bullmore ET, Smith M, Christensen S, Nathan PJ, Rockstroh B, Bassett DS (2009) Reproducibility of graph metrics of human brain functional networks. NeuroImage 47(4):1460–1468. PMID: 19463959
395. Hayasaka S, Laurienti PJ (2010) Comparison of characteristics between region-and voxel-based network analyses in resting-state fMRI data. NeuroImage 50(2):499–508. PMID: 20026219
396. Kaiser M, Martin R, Andras P, Young MP (2007) Simulation of robustness against lesions of cortical networks. Eur J Neurosci 25(10):3185–3192.
397. Seeley WW, Crawford RK, Zhou J, Miller BL, Greicius MD (2009) Neurodegenerative diseases target large-scale human brain networks. Neuron 62(1):42–52. PMID: 19376066
398. Sanz-Arigita EJ, Schoonheim MM, Damoiseaux JS, Rombouts SARB, Maris E, Barkhof F, Scheltens P, Stam CJ (2010) Loss of 'small-world' networks in Alzheimer's disease: graph analysis of FMRI resting-state functional connectivity. PloS One 5(11):e13788. PMID: 21072180
399. Sporns O, Kötter R (2004) Motifs in brain networks. PLoS Biol 2(11):e369
400. Alon U (2007) Network motifs: theory and experimental approaches. Nat Rev Genet 8(6):450–461
401. Cook SA (1971) The complexity of theorem-proving procedures. In: Proceedings of the third annual ACM symposium on theory of computing, STOC'71, Shaker Heights. ACM, New York, pp 151–158
402. Hagmann P, Cammoun L, Gigandet X, Meuli R, Honey CJ, Wedeen VJ, Sporns O (2008) Mapping the structural core of human cerebral cortex. PLoS Biol 6:e159
403. Artzy-Randrup Y, Fleishman SJ, Ben-Tal N, Stone L (2004) Comment on network motifs: simple building blocks of complex networks and superfamilies of evolved and designed networks. Science 305(5687):1107–1107. PMID: 15326338
404. Ingram PJ, Stumpf MP, Stark J (2006) Network motifs: structure does not determine function. BMC Genomics 7(1):108. PMID: 16677373
405. Knabe JF, Nehaniv CL, Schilstra MJ (2008) Do motifs reflect evolved function?–no convergent evolution of genetic regulatory network subgraph topologies. Bio Syst 94(1-2):68–74. PMID: 18611431
406. Hell P, Nesetril J (2004) Graphs and homomorphisms. Oxford University Press, Oxford/New York

407. Widom B, Rowlinson JS (1970) New model for the study of Liquid-Vapor phase transitions. J Chem Phys 52(4):1670–1684
408. Dyer M, Greenhill C (1999) The complexity of counting graph homomorphisms. In: 11th ACM/SIAM symposium on discrete algorithms. Press, San Francisco, pp 246–255
409. Strogatz SH (2001) Exploring complex networks. Nature 410(6825):268–276
410. Newman MEJ (2000) Models of the small world. J Stat Phys 101(3-4):819–841
411. Ehresmann A (1997) Colimits in free categories. Diagrammes 37:3–12
412. Mahalanobis PC (1936) On the generalised distance in statistics. Proc Natl Inst Sci India 2(1):49–55. Retrieved 2012-05-03
413. Girvan M, Newman MEJ (2002) Community structure in social and biological networks. Proc Natl Acad Sci 99(12):7821–7826. PMID: 12060727
414. Newman MEJ (2006) Modularity and community structure in networks. Proc Natl Acad Sci 103(23):8577–8582. PMID: 16723398
415. Onnela JP, Saramäki J, Kertész J, Kaski K (2005) Intensity and coherence of motifs in weighted complex networks. Phys Rev E Stat Nonlinear Soft Matter Phys 71(6 Pt 2):065103. PMID: 16089800
416. Song C, Havlin S, Makse HA (2005) Self-similarity of complex networks. Nature 433(7024):392–395
417. Kant I, Weigelt M, Müller FM (2007) Critique of pure reason. Penguin, London/New York
418. Polger T (1949) Philosophy of mathematics and natural science. Princeton University Press, Princeton
419. Hull CL (1952) A behaviour system: an introduction to behavior theory concerning the individual organism. Yale University Press, New Haven
420. O'Keefe J, Nadel L (1978) The hippocampus as a cognitive map. Oxford University Press, New York
421. O'Keefe J, Dostrovsky J (1971) The hippocampus as a spatial map. Preliminary evidence from unit activity in the freely-moving rat. Brain Res 34(1):171–175. PMID: 5124915
422. Milford MJ (2008) Robot navigation from nature. Springer, Berlin
423. Redish A (2001) The hippocampal debate: are we asking the right questions? Behav Brain Res 127(935):81–98
424. Tolman E (1948) Cognitive maps in rats and men. Psychol Rev 55(4):189–208
425. Muller R, Kubie J, Ranck J (1987) Spatial firing patterns of hippocampal complex-spike cells in a fixed environment. J Neurosci 7:1935–1950
426. Fyhn M, Molden S, Witter M, Moser E, Moser MB (2004) Spatial representation in the entorhinal cortex. Science 305:1258–1264
427. Hafting T, Fyhn M, Molden S, Moser MB, Moser E (2005) Microstructure of a spatial map in the entorhinal cortex. Nature 436:801–806
428. Moser E, Kropff E, Moser M (2008) Place cells, grid cells, and the brain's spatial representation system. Annu Rev Neurosci 31:69–89
429. Solstad T, Moser E, Einevoll G (2006) From grid cells to place cells: a mathematical model. Hippocampus 16:1026–1031
430. Zilli E, Yoshida M, Tahvildari B, Giocomo L, Hasselmo M (2009) Evaluation of the oscillatory interference model of grid cell firing through analysis and measured period variance of some biological oscillator. PLoS Comput Biol 5(11):e1000573
431. Ranck J (1984) Head-direction cells in the deep cell layers of dorsal presubiculum in freely moving rats. Soc Neurosci Abstr 10:599
432. Taube JS (1995) Head direction cells recorded in the anterior thalamic nuclei of freely moving rats. J Neurosci 15(1):70–86. PMID: 7823153
433. Mizumori SJ, Williams JD (1993) Directionally selective mnemonic properties of neurons in the lateral dorsal nucleus of the thalamus of rats. J Neurosci 13(9):4015–4028. PMID: 8366357
434. Wiener SI (1993) Spatial and behavioral correlates of striatal neurons in rats performing a self-initiated navigation task. J Neurosci 13(9):3802–3817. PMID: 8366346

435. O'Keefe J (1991) An allocentric spatial model for the hippocampal cognitive map. Hippocampus 1(3):230–235. PMID: 1669295
436. McNaughton BL, Barnes CA, Gerrard JL, Gothard K, Jung MW, Knierim JJ, Kudrimoti H, Qin Y, Skaggs WE, Suster M, Weaver KL (1996) Deciphering the hippocampal polyglot: the hippocampus as a path integration system. J Exp Biol 199(Pt 1):173–185. PMID: 8576689
437. Eichenbaum H (2002) The cognitive neuroscience of memory: an introduction. Oxford University Press, Oxford/New York
438. Adámek J, Gumm HP, Trnková V (2009) Presentation of set functors: a coalgebraic perspective. J Log Comput, 20(5), 991–1015
439. Touretzky D, Redish A (1996) Theory of rodent navigation based on interacting representations of space. Hippocampus 6(3):247–270
440. Burgess N, Recce M, O'Keefe J (1994) A model of hippocampal function. Neural Netw 7:1065–1081
441. Arleo A, Gerstner W (2000) Spatial cognition and neuro-mimetic navigation: a model of hippocampal place cell activity. Biol Cybern 83:287–299
442. Jung M, Wiener S, McNaughton B (1994) Comparison of spatial firing characteristics of units in dorsal and ventral hippocampus of the rat. J Neurosci 14:7347–56
443. Kjelstrup K et al (2007) Very large place fields at the ventral pole of the hippocampal CA3 area. Soc Neurosci Abstr 33(93):1
444. Wilson RA, Keil FC (eds) (1999) The MIT encyclopedia of the cognitive sciences. MIT, Cambridge
445. Scoville W, Milner B (1957) Loss of recent memory after bilateral hippocampal lesions. J Neurol Neurosurg Psychiatry 20:11–21
446. Jones EG, Mendell LM (1999) Assessing the decade of the brain. Science (New York, NY) 284(5415):739. PMID: 10336393
447. Raichle ME (2010) Two views of brain function. Trends Cogn Sci 14(4):180–190. PMID: 20206576
448. Cohen MA, Grossberg S (1987) Absolute stability of global pattern formation and parallel memory storage by competitive neural networks. In: Grossberg S (ed) Advances in psychology, vol 42. North-Holland, Amsterdam, pp 288–308

Index

J. Gómez-Ramirez, *A New Foundation for Representation in Cognitive and Brain Science*, Springer Series in Cognitive and Neural Systems 7, DOI 10.1007/978-94-007-7738-5,

MIX
Papier aus verantwortungsvollen Quellen
Paper from responsible sources
FSC® C105338

If you have any concerns about our products,
you can contact us on
ProductSafety@springernature.com

In case Publisher is established outside the EU,
the EU authorized representative is:
Springer Nature Customer Service Center GmbH
Europaplatz 3, 69115 Heidelberg, Germany

Printed by Libri Plureos GmbH
in Hamburg, Germany